THE LONG PATROL:

CSS H.L. HUNLEY, CHARLESTON, AND THE CIVIL WAR 1863-2004

by

Michael J. Kozlowski

Dedication

...   For Melissa, who taught me to believe in Second Miracles

...   For Michael, who Understands

...   With Respect to Dr. Clive Cussler, the man who answered the Light

...   And most of all, to Twenty-two brave men who changed history.

Copyright Notice

Copyright © 2013 Lion Publications Inc, 22 Commerce Road, Newtown, Connecticut 06470. ISBN 978-1-939335-14-2 No part of this original work may be reproduced or transmitted in any form or by any means, electronic or mechanical including photocopying, recording or by any information and retrieval system without permission in writing from the publisher

TABLE OF CONTENTS

I: "LORD GOD, OUR POWER EVERMORE…" 1

II: "WHOSE ARM DOTH REACH THE OCEAN FLOOR…" . 21

III: "DIVE WITH OUR MEN BENEATH THE SEA…" 47

IV: "…TRAVERSE THE DEPTHS PROTECTIVELY…" 66

V: "…O HEAR US WHEN WE PRAY, AND KEEP…" 108

VI: "…THEM SAFE FROM PERIL IN THE DEEP." 190

VII: VALEDICTORY 222

ACKNOWLEDGEMENTS 256

REFERENCES AND BIBLIOGRAPHY 258

APPENDICES 262

APPENDIX 1: CSS PIONEER 262

APPENDIX 2: INTELLIGENT WHALE 264

APPENDIX 3: PIONEER II 266

APPENDIX 4: THE GALENA LETTER 268

APPENDIX 5: SUBMARINE STATISTICS 268

THE AUTHOR

Mike Kozlowski was born in Cleveland, Ohio, in 1960. He enlisted in the United States Air Force in 1978, beginning a career that spanned three decades and saw the rebuilding of US forces after Vietnam, the collapse of the Soviet Union, and Operation Desert Storm. He served as a Munitions Specialist with some of the USAF's most elite units, including the legendary Strategic Air Command and the 1st, 8th, and 20th Fighter Wings. He is a graduate of the USAF's Combat Ammunition Course and was selected three times to serve as a Flightline Supervisor for the USAF's famous Red Flag exercises in Nevada. Mike has been awarded the USAF Senior Maintenance Badge, Missileman's Badge, the Air Force Commendation Medal with two Oak Leaf Clusters, the USAF Small Arms Expert Ribbon, and the Southwest Asia Service Ribbon. He holds a degree in Munitions Systems Technology, and a Bachelor's Degree with Honors from the University Of South Carolina. He resides in Columbia, South Carolina with his wife Melissa and a slightly odd cat named Eek! .

I: "LORD GOD, OUR POWER EVERMORE…"

Charleston calls it the 'Antique District', and that's not far off – though certainly not in the sense they'd like you to take it. It is a semi-depopulated area of light industry and empty buildings, with empty, rattletrap houses poking up from islands of green/brown scrub with the sounds of traffic ricocheting off the weather-beaten brick and wood from nearby Interstate 26.

This being America, though, it is almost axiomatic that there are small gems to be found in places like the Antique District and this is no exception. Off of Meeting Street is a small cluster of cemeteries – once pleasantly out in the country from old Charleston, then subsumed in the growing city and finally bypassed. Most of them are spare, direct spaces whose condition ranges from neatly kept and manicured to flat out derelict, the surviving stones emerging from overgrown weeds like shipwrecks on a hostile shore. The worst ones have a certain grim despair about them; as if the last tangible reminders of their residents are finally about to disappear into eternity and we have no wish to consider that this may be our ultimate fate someday. We shall walk past them with heads held high, literally whistling past the graveyard.

It is the others that draw our attention, even curiosity. There is Bethany – neat, laid out in a precise, severe grid that is filled with Teutonic names dating back to the 1820s. More than a few of the stones are engraved in German, but whether or not one can understand it the traditional format is familiar enough that one can get the gist of it – *Geborn 1817, Gestorben 1861*. There is the Friendly Union, a much smaller space than Bethany, but filled with more familiar Southern names.

There are two or three others, then you come to the jewel in the crown – Magnolia. Officially, it dates back to 1849, but there are some small, almost illegible stones that seem to indicate a few years earlier than that – far enough back that their owners were born subjects of His Majesty King George The Third.

"The jewel in the crown – Magnolia…" Magnolia Cemetery, Charleston, S.C. (Photo courtesy Theodore Leverett)

It is a fairly good-sized place, larger than the rest put together. It is beautifully kept, with Spanish moss and palmettos providing cool, comforting shade. A string of small ponds winds through it, running past dirt paths whose half buried fragments of red brick hint at a time long ago when horse-drawn hearses clip-clopped at a stately pace towards one of the hundreds of discreetly walled family plots, while men in black and women in veiled widow's weeds trudge behind it. The plots draw your attention – many with a low wall or railing running around them, low enough to comfortably step over without breaking your stride. The earliest ones often have long, almost poetic descriptions of the lost one along with their accomplishments, but as time goes on, the inscriptions tend to dwindle down to "Dear Father" or "Beloved Mother", then to a simple set of dates. Several stones proclaim their owners as former Governors of the Palmetto State, as well as Senators, Attorney Generals, Solicitor Generals, and other lesser political lights.

Many stones tell heartbreaking stories without saying a word. There are many smaller stones from the mid-1800s that have a lamb carved into them. Without exception, these have birth and death dates that are separated by only a few years, perhaps even just a few months or days. Some, with poignant horror, have no dates at all. One tells of the death of a young woman "In The Time Of The Plague", one of the yellow fever epidemics that periodically ravaged Charleston. If you take the time to look, you will often see clusters of stones from those dates. The earliest ones are ornately carved, but as the dates of the Plague Time crawl by, the stones become smaller and plainer, until finally they are no more than small stone loaves with initials and a final date engraved upon them. Mass death has a speed and haste all its own that sometimes gives us little time for the formalities we hope for, and the Plague Times did not always permit them.

There are other plain stones there as well, fields of them that simply say, "UNKNOWN 1863" or "UNKNOWN 1864". These are the gravesites of hundreds of Confederate soldiers and a handful of sailors, who rest at Magnolia. Most of them actually died of disease of one sort or another, one of the epidemics that periodically swept the camps and worsened by poor nutrition and exposure. We tend to forget that in the Civil War more than half of the casualties in that awful butcher's exercise were from non-combat causes, and we shake our heads in sadness at the futility of it all.

Some of the warriors who lie there quiet and still have been identified through the efforts of the Sons Of Confederate Veterans, who have dedicated themselves to finding and giving names to these lost souls who died for a cause that did not outlive them by much. Their stones are shiny new marble, carved identically to those that the Federal government gives to retired veterans and those lost in action. They are dignified and noble, and the contrast to those little stones with "UNKNOWN" is all the greater.

It is at the far rear of the cemetery where a small green island encircled by the remains of a brick path looks out over the marshes that lead out to the Wando River. The most modern thing in sight from there are the towers of the Arthur Ravenel Bridge a few miles away, while trees hide the towering stack of a manufacturing plant of some sort. The most recent stones there have dates into the middle and late twentieth century, ending family lines that other stones show dating back before there was a United States Of America. In that island now lie twenty-one men, America's first submarine crews. The last of them were laid to rest in April, in the Year of Our Lord Two Thousand and

Four, ending a journey that covered about fifty miles and nearly a century and a half.

The long journey to Magnolia starts not in Charleston, but far away in New Orleans in the summer of 1861.

America had gone slightly mad and was at war with itself. The first shots were fired in Charleston on April 12th under the command of Pierre Gustave Toutant Beauregard, a flamboyant Cajun from the Big Easy.

Beauregard's life and career up to that point were a good example of how badly things had fractured. He'd graduated from West Point in 1838, and after a solid career he decided to try his hometown's politics for a while. He later returned to the Army and became Superintendent of West Point...for exactly five days in early 1861. He made no secret of his support for the South in the rapidly approaching conflict and when the Union started coming apart after Lincoln's election, the senior Army leadership wanted him out of the Point, no doubt thinking of another former commander there named Benedict Arnold. For his part, Beauregard resigned his commission and headed south in February where he jumped from Captain, US Army, to Brigadier General, Confederate States Army, and was sent to Charleston, where he oversaw the first shots directed at Fort Sumter.

In truth though the war had started long, long before that – and one could make the case that the seeds had been planted back in the late 1700s, when the Constitution was ratified but the Several States still had a great deal of autonomy, to an extent we wouldn't recognize now. Slavery was the match to the fuse, but the kegs of gunpowder that it led to had been piling for decades.

The states that formed the Confederacy felt that their history, their traditions, their laws, and their customs were being trampled on, and they demanded – rightfully so, in their eyes – either protection from those who would change them at cold steel's point, or the option of leaving the whole rotten Union altogether. On the other side, abolitionists and advocates of a strong Federal government considered their wayward brethren at best arrogant rebels and at worst fiendish, genocidal traitors.

The Long Patrol

The shooting started long before that first cannonball arced up over Fort Sumter – the homicidal rampages of both pro- and anti-slavery men in the Kansas Territory were bells in the night, warning both sides of a coming storm. John Brown – whose Body, moldering in a New York grave, was sung of so often during those days – tried to start a slave rebellion by taking the Federal arsenal at Harper's Ferry, VA, but succeeded only in killing one free black and getting himself surrounded, captured, tried, and executed.

The middle ground had long since vanished; all that remained was a growing chasm that seemed to call both sides to the brink. The election of Abraham Lincoln as President in November 1860 convinced many Southern leaders that there was no further chance at reconciliation, though in all honesty they may not have needed much convincing. South Carolina, long a guiding star of Southern politics and political thought, led the way out on December 20th, 1860 and the other states fell like dominos thereafter.

Lincoln tried to be conciliatory, speaking of 'our better angels' in an attempt to calm the Confederacy down and stop the Union's disintegration. But he could only go so far. For Abraham Lincoln, preservation of the Union – by whatever means necessary – was the *sine na quon* of his Presidency. Through masterful political strategy, he handed the Confederacy a no-win choice: allow the garrison at Charleston's Fort Sumter to be resupplied and make a mockery of their claims to sovereignty, or use military force to stop the supply ship, thereby taking on the onus of having fired the first shot.

On April 12, 1861, the Better Angels lost. One battery's staff allowed a rabid Secessionist named Edmund Ruffin to pull the lanyard that fired the first round, one that went racing skywards over Charleston Harbor, silencing the crowds that had gathered on the Battery to watch the excitement until it burst with a dull *thud* in the unfinished masonry pile that was Fort Sumter. The crowd cheered a nonstop bombardment that lasted three days until the Sumter garrison surrendered. For all the violence of the assault, there were only five Federal casualties – a single soldier killed during the surrender ceremony when a saluting gun burst, with one more dying of his injuries later, and a three more severely injured - and no Confederate casualties at all.

Given what we have seen of war today it seems odd, even disturbing, that there should have been crowds cheering as Sumter was beaten into submission. But remember that no one was killed during the actual battle, and after all, Confederate leaders asked, why should there even *be* a war now? They had made their point, they had taken property that was rightfully theirs, and they had defended their homes. Now Lincoln would have to deal with them as equals and acknowledge their independence, and if he wanted to fight them again they'd give him more of the same.

On the other hand, many a Federal politician and soldier felt that the attack on Sumter had been a stab in the back on an unprepared garrison in an unfinished fortress. Up against the muskets and sabers of regular US Army troops and the rapidly massing volunteers, there would be no possible way the infant Confederacy could possibly survive for more than a few months. Either way, neither side expected a long war, and more than one confident speaker on both sides offered to drink all the blood that would be spilled.

Rarely in American history would there ever be other miscalculations so disastrous.

By late summer, both sides had stepped back and taken the measure of one another. The Battle of First Bull Run – or First Manassas, to those below the Mason-Dixon line - on July 27th had ended with both armies *hors d'combat*. The Federals were routed so badly that some units did not stop running until they reached the defenses of Washington, forty-odd miles away, while the victorious Confederates were so badly disorganized in the aftermath of their victory that they could not have followed up on it if they'd had to. In the meantime, between Bull Run and a rapidly mounting number of small unit skirmishes, the blood spilled was becoming enough to drown a division.

Lincoln's volunteers would take time and effort to train, while the Confederacy was still desperately trying to organize itself. On the other hand, the Confederacy felt it could at least take some comfort in the fact that the Union Army would be beyond the ability to conquer them for some time to come. It was yet another miscalculation, for the greatest threat to the Confederacy at that time was not from the land but from the sea. On April 19th, Lincoln had signed a decree ordering a blockade of Confederate ports along the Atlantic and Gulf seacoasts.

"...Confederate naval officers were few in number but magnificent in talent..." Seated, left to right, Admiral Franklin Buchanan, Captain Josiah Tattnall, Commander Matthew Maury. Standing left to right, Captain George Hollins, Rear Admiral Raphael Semmes, and Secretary of the Navy Stephen Mallory. (Photo courtesy of the United States Naval Historical Center, Washington, DC.)

Now, Confederate naval officers were few in number but magnificent in talent. The courtly swashbuckler Raphael Semmes, rock-solid old Franklin Buchanan, the brilliant oceanographer and technical specialist Matthew Maury – they and others gave the Confederate States Navy a small but solid core to build from. And their opinions on Lincoln's actions were unanimous.

First, a blockade was ridiculous from a political view, for international law said that a blockade could only be used against an enemy nation – and acknowledging the CSA as an enemy nation came perilously close to acknowledging its independence in regards to other nations.[1]

Secondly, a blockade was considered almost insane from a tactical viewpoint. The US Navy of 1861 was a small force mostly intended to keep American commerce protected and show the flag on occasion – at least those vessels capable of moving out to sea. Of ninety-odd warships in commission, a fairly high number of the USN's ships – perhaps as high as half of them - weren't even capable of leaving dockside, being what was called 'in ordinary'. That deceptive little phrase meant that they had been simply tied up long ago and left there with small caretaker crews or no one at all and quietly ignored to settle a little deeper in the water every day.

Among them was the USS *Pennsylvania*, one of the first 'ships of the line' in the US Navy. It had taken *twenty-six years* to build her, and her only cruise was a quick trials run from Philadelphia to Norfolk, where she was now but a glorified dormitory and warehouse. There was also the USS *United States*, the legendary 'Old Waggon', sister to USS *Constitution* and victor against the French and the British. Her glory though had long since fled and she was rotting at dockside in Norfolk.

The remaining ships were scattered all over the world's oceans, and it would take months for them to get back to US waters. Even then, there would be no way to deploy sufficient numbers to cover thirty-five hundred miles of coastline.

The Confederacy's naval experts were unanimous: it could not be done without a shipbuilding program that would take years and bankrupt the Union. Needless to say, it was a shock when ships started showing up outside Confederate ports within a few weeks. The Union had simply started buying merchant ships, ships on the ways, and anything else they could get their hands on that could float and move. The CSN's officers,

---

[1] E. Milby Burton points out that a much safer course of action would have been to simply announce that certain ports were closed due to a state of insurrection, and that the US Navy was simply there to enforce that closure. Calling it a blockade was one of the rare diplomatic missteps that can be laid directly to the Lincoln Administration – Burton, *The Siege of Charleston 1861-1865*, University of South Carolina Press, 1970, pg 242

schooled in the traditions of Nelson and Drake, had forgotten that the world was changing – and so were the rules.

The Confederate States Navy, for all its glory, would never approach even a fraction of the US Navy's strength. They had captured several major warships at Norfolk, but they were either burned beyond recognition – like *Pennsylvania* was – or derelict beyond recovery – like *United States* – or meant for other fates, like the steam frigate *Merrimac*, burned to the waterline where she sat. In any event, most of those ships could never be brought into service for the CSN. Manpower was problematical from the first day of the war, as most of the USN's strength had come from the North. There was enough to keep the ships they had manned, but just barely, and the manpower they had was often unskilled.

What we remember from the CSN's war were the raiders; fast sleek ships built in Europe and turned into lethal weapons. They were ships like CSS *Alabama*, built in Liverpool and the terror of the US merchant fleet, or the CSS *Shenandoah*, which single-handedly decimated the Union whaling fleet - some of it as late as June 1865, two months after the war ended. There was the CSS *Florida*, commanded by a rakish devil named John Maffitt, which took fifty-five Union merchantmen and was only caught and captured near the end of the war in Brazil by a USN force that violated several major tenets of international law. For all their successes though, they were not much more than converted fast merchantmen.

The CSN did commission two ironclad rams ostensibly built for the Pasha of Egypt that would have been the equal of anything the USN could have put to sea and could very well locally broken the blockade line – once - if they worked together.[2] After that, they would have probably rotted at their moorings, trapped by a very upset United States Navy, or doomed to a fast, glorious end as they tried to shoot their way out again. But they were still two of the world's most advanced warships at the time, and they were a threat, so on this matter the

---

[2] Bruce Catton, *The American Heritage Picture History Of The Civil War*, Random House, New York, 1960, pg. 260.

Union government stood its ground. Charles Francis Adams, Ambassador to the Court of Saint James, delivered a note to Her Majesty's Government that stated the facts quite bluntly:

> *"...It would be superfluous in me to point out to your Lordship that this is war..."*
>
> *Charles Francis Adams*
>
> *September 5, 1863*

The British knew it too so they stopped the rams from ever getting to the Confederacy - bending their own and international law rather substantially in the process. That finished what little chance the CSN ever had of taking on the US Navy on anything like an equal footing. The Confederacy, however, still had the problem of dealing with the blockade.

The best known and ultimately the most effective of the solutions was that romantic and daring character known as the blockade-runner. They were some of the fastest ships in the world at the time, often screw or sidewheel steamers painted a haze gray to reduce their visibility and burning low-smoke coal, their skippers gutsy, highly skilled men who could sneak their charges past the Federals at night or in bad weather – and make a fortune if they could make it through the blockade. At first many did, and a false sense of confidence rose throughout the South, especially in Charleston. If this was the best the Yankees could do, they might just have a chance.

They didn't.

The blockade took some time to get its sea legs, so to speak, but once it did the crews dug in with a vengeance, motivated by both the opportunity to help strangle the rebellion and the chance to make literal fortunes in prize money. As the blockade tightened, more and more runners brought in luxury goods instead of necessities – priceless in the Confederacy, and still worth a king's ransom in the North. Officially, Confederate law required runners to bring in military supplies and other requirements, but no one seems to have made much of an effort to enforce it except *in extremis*. And in the end the blockade never completely stopped the runners; they didn't stop until the last of the Confederate ports was occupied in February of 1865. There was though another option open to the Confederacy, one with far greater risk to its practitioners: the privateers.

It was a fairly time-honored custom, most popular in the 17th and early 18th centuries: one would petition one's government for a letter of marque, which authorized you to go out and hunt down and capture enemy shipping. The government, of course, got a cut, it wouldn't have been the government otherwise, but you got to keep the rest. Assuming you made it out, captured enough shipping to make it worth it, and then got back in through the blockade again, it could be an extremely lucrative vocation. Privateering commissions were handed out to individuals and syndicates alike, with fast armed merchants and homebuilt iron rams.

There was one serious drawback to this particular vocation – most nations regarded privateering as thoroughly, utterly, and completely illegal. By 1861 it had been pretty much outlawed by the world's leading nations except the United States[3]. But as a rule, anyone the USN caught privateering would be considered a pirate, and treated accordingly. That, as a matter of routine, consisted of a quick, fair trial followed by a hanging. But as we have since discovered again and again, governments under threat of defeat tend to find reasons to take actions that other nations have long since banned, and the Confederacy was no different.

It should be pointed out that the Confederate Constitution - Section 8, Article 11 - authorized the Confederate Congress to 'grant letters of marque and reprisal', and as such made it at least theoretically legal for privateers to ply their trade in the service of the CSA. Potential privateers could also take comfort in the fact (or desperately convince themselves) that since the CSA was not a signatory to the Treaty of Paris - didn't even exist then, for that matter - any law governing their treatment if captured was at best murky.

---

[3] The 1856 Treaty of Paris outlawed privateering by the United Kingdom and France, the two largest navies of the time. However, the US was not a signatory. During the Civil War and the Spanish-American War, the US Government policy was that no privateering would be allowed but officially, the Constitution still authorizes letters of marque – government permission to become a privateer. - MJK

CSS H.L. Hunley

But in the end, it really was all about money. Those who wanted to go after Yankee shipping the old fashioned way - in uniform - would do so. But for the others...the potential for delivering a blow against the enemy and the sheer amount of money involved tended to bring out an odd assortment of patriots, pirates, and other interested parties who were attracted by the fact that they as private citizens could get in a shot at the Union *and* make a profit. One of them was a Tennessee preacher named Franklin Smith, who had some unique ideas about exactly how to break the blockade.

For some reason, clergymen and doctors tend to show up in places we don't expect them in the Civil War. There was the Reverend Leonidas Polk - an Episcopal bishop, no less - who had graduated from the Point but later left the service. He returned after Secession and was immediately appointed a Major General in the CSA. He never quite lived up to his early billing, but still hung gamely on until he went down before Atlanta, desperately trying to stop William Tecumseh Sherman from splitting the Confederacy asunder. There was the Reverend William Scott in California, whose support of states' rights eventually wound up with him fleeing the Golden State for Europe after he voted against a series of pro-union resolutions and, not incidentally, starting a riot in the process.[4] Doctor Richard Gatling, MD, felt that designing a weapon that could fire hundreds of rounds a minute would make war so appalling that it would no longer happen. The type of gun that still bears his name today has instead become a synonym for sheer ferocity.

There was one other healer whose efforts have a grimly 21st-century ring to them: Doctor Luke Pryor Blackburn, M.D., of Kentucky by way of Halifax, Canada. He may - stress the 'may', as there is still some doubt as to whether or not he actually had this in mind - tried to start a yellow fever epidemic in New York City by shipping clothes and blankets covered in bloody vomit to that benighted city. As the time, it was believed that yellow fever could be spread by such means (the vector theory of disease hadn't yet been put forward) and the possibility

---

[4] http://www.militarymuseum.org/CAandCW2.html

was taken seriously enough that Her Majesty's Government confiscated the shipment of clothes, and only narrowly missed confiscating Doctor Blackburn himself. Some sources suggest that Blackburn was also trying the same homicidal experiment with smallpox-infected blankets...and *that* might have worked.

The Center for Disease Control says in their own briefings that it is rare but *possible* for smallpox to be spread by contact with infected "bedclothes, linens, or blankets"[5], and in the filthy, crowded slums of New York all it would have taken was one case of smallpox to spin out of control, and it would have spread like wildfire. A full-blown smallpox epidemic loose in New York City would have killed tens of thousands and severely dislocated the Federal economy - possibly to the point where the Lincoln Administration might have had to look hard at continuing the war.

Add to that the fact that New York's mayor, Fernando Wood, was publicly entertaining the idea of his own city seceding to become a 'free city', making a fortune from both sides. New York City was the Union's weak link, and the Confederacy knew it. If Blackburn was trying to hurt the Union through biological warfare, Washington itself wouldn't have been as attractive a target as the Big Apple.

Blackburn stayed one step ahead of the law from several nations before finally being brought to trial in Canada in 1865. He never took the stand in his own defense; for that matter he never made any kind of statement on the subject, written or otherwise, but he was ultimately acquitted fair and square. Returning to Kentucky, he helped deal with two yellow fever epidemics and made such a good impression that he was elected Governor in 1878, and died regarded as a hero in 1887.

It should, in honesty and fairness, be pointed out that Blackburn's primary accuser - a Confederate agent in Canada named Godfrey Hyams - may have been seriously overstating the case against the good Doctor, going so far as to accuse him

---

[5] 'History and Epidemiology of Smallpox Eradication', Slide 19, from the training course titled *"Smallpox: Disease, Prevention, and Intervention"* Centers for Disease Control and Prevention, Atlanta, GA. www.bt.cdc.gov/agent/smallpox/training/overview,

of planning to send President Lincoln contaminated clothing and making that charge just two days before Lincoln's assassination. In the fever-pitch atmosphere of those awful days, between the collapsing Rebels and Lincoln's murder, no charge of potential treason on the part of the Confederacy or against the martyred President was too wild to be believed, and the effort to get Blackburn would have easily snowballed from there.

On the other hand, the British authorities in Bermuda received information independent of Hyams' claims that led them to that stockpile of contaminated garments and blankets. Edward Steers, in his book *Blood on the Moon: The Assassination of Abraham Lincoln,* puts the case plainly: there may not have been enough evidence to prove in court that Blackburn was planning strategic biological warfare, but there is plenty of evidence to show Blackburn was at least aware of a plot...and so were some of the highest-ranking members of the Confederate Government.[6]

But let us leave this unpleasantness and return to the subject of Reverend Smith.

According to Mark Ragan in his masterpiece *The Hunley: Submarines, Sacrifice and Success In The Civil War*, the Reverend Smith was a chemist and inventor of exceptional ability who was quite well known for his work throughout the South, and he wrote a letter to a local paper suggesting a unique solution to the blockade: the submarine.[7]

Reverend Smith was far from the first person to come up with it – Leonardo DaVinci designed one centuries before, though it wasn't terribly practical. A Dutchman named Cornelius Van Drebbel had designed and built a sealed rowboat that was actually able to submerge, though it did so more through brute force than any sophisticated ballast system.

During the Revolutionary War, David Bushnell built a

---

[6] Steers, Edward, *Blood on the Moon: The Assassination of Abraham Lincoln*, The University Press of Kentucky, Lexington, KY, 2000, pg 47 - 51.

[7] Mark Ragan, *The Hunley: Submarines, Sacrifice & Success In The Civil War*, Narwhal Press, Charleston, 1999, pg. 18.

surprisingly workable submarine from two barrels called the *Turtle*, which actually got close enough to the British ship of the line HMS *Eagle* to actually try and attach its mine – allegedly to be defeated when the *Turtle's* auger couldn't puncture the *Eagle's* copper-covered bottom. In fact, *Eagle* wasn't coppered at the time and (prophetically) the failure was probably due to exhaustion and hypothermia. The legendary inventor Robert Fulton, just half a century before the Civil War, had proposed a submarine to the British first (who turned it down because they found it ungentlemanly) and then to Napoleon. His Majesty found the idea interesting, but certain military reversals kept him from ever putting it into use.

However, by 1861, the technical means to build a functional submarine – construction techniques, ballast systems, and even propulsion – were finally advanced far enough to create the possibility of a practical submarine, and Reverend Smith was able to grasp that. His letter pointed out that an effective submarine would have to be cigar-shaped and streamlined, without external protrusions like rivet heads. It would need a central propeller, and an accurate means of determining its depth beneath the surface. All of these things were now technically possible, and if such a vessel could be built, it would be a highly effective weapon against the blockade. Insofar as they go, those are still requirements for the lethal nuclear attack submarines of today. The question though, was whether or not the Confederacy could build one.

Bushnell's Turtle. Source: U.S. Navy

The Confederacy was overwhelmingly an agricultural nation with an almost non-existent industrial base save for a few very small foundries and iron works. There were no armories capable of making cannon, and only one facility – the legendary Tredegar Iron Works in Richmond – capable of making the kind of rolled plate that would be needed for a submarine, and its entire output was spoken for into the foreseeable future. There were shipyards but they were small, mostly intended for repair and refit, and none of them had the ability to build anything made of iron. On the other hand, the North seemed to have advantages that could never be overcome: access to infinite amounts of raw materials, entire *cities* dedicated to steel production, shipyards that could turn out anything from rowboats to ironclad ships of the line, and the ability to arm them with state-of-the-art firepower and enough sailors to man a dozen other navies. On the face of it the Reverend Smith's idea, though clearly stated and technically accurate, was one of those things that could never be brought to fruition.

"the legendary Tredegar Iron Works in Richmond" Source: National Archives.

Which brings us back to New Orleans.

Horace Lawson Hunley looks out at us over a century and a half from a battered Daguerreotype with a look of – well, it's tough to say exactly what it is. It's not a grin, but it is a look of confidence and assurance that gives us a remarkable insight into what kind of man he was. His hair and beard are well trimmed, and he has an air of quiet success about him, as if even this far away in time, he is still able to assure us of his knowledge and abilities. He had every reason to look confident – he had been exceptionally good at everything he'd done up to this point.

Hunley was born in Sumter County, Tennessee, on

December 29th, 1823, the son of a cotton broker who had moved to New Orleans to be closer to the action. Things - as they so often do in situations like that - did not necessarily work out in favor of the Hunleys, and went from bad to worse when his father died in 1834. Without funds to move back to Tennessee the Hunleys stayed on but, in a almost Dickensian twist, Louise Hunley met and married a wealthy planter from New Jersey. The planter not only loved Louise dearly, but took care of young Horace as well, using his business and society connections to set Hunley up right - starting with a law degree from what is now present-day Tulane University in 1849, followed by highly successful business ventures.

Horace Lawson Hunley. Courtesy Patriot's Point Naval Museum.

Hunley read voraciously, and from early in life he had read the lives of heroes and successful men, determined that the poverty he had once known would never happen again - and that his name would ring through the centuries as well.[8]

Horace Hunley was one of those men who so often appear in the 18th and 19th centuries, and whom we sometimes scoff at today – brilliant and highly successful at a number of things that in our world would require a lifetime's devotion to be merely competent at even one. He entered politics early, serving in that bear's den known as the Louisiana Legislature. He became a very well known lawyer in the Big Easy, as well as a successful planter and businessman, all of those things before the age of forty - *quel homme!*

Another high point came in 1857, when he was appointed to

---

[8] "Horace L. Hunley", *American National Biography, Supplement 2,* Mark C. Carnes, editor, Oxford University Press, New York, New York, 1998

CSS H.L. Hunley

a Customs House post - not to be misconstrued as a sign of support for the Federal Government, to be sure, but rather simply as a reward for Services Rendered. The pay was good, plus he was allowed to live in the then-unfinished Customs House (and it would stay unfinished through hurricane, corruption, and civil war until 1881) at the corner of North Peters and Canal. When the Union began to disintegrate, Horace Hunley then was in a remarkable position to see a great many things that the average Confederate citizen could not - or in the case of many of those who fought hardest for secession, would not. One does not achieve the things Hunley had without being very realistic.

The Confederate blockade runner Ella and Annie. She was captured by Union forces in 1863, then purchased by the Union Navy as the USS Malvern. Post-war she reverted to civilian service and was wrecked in 1895. Source: U.S. Navy.

First, New Orleans was the center of the cotton industry - the staple crop of the South, and the Confederacy's ace in the hole. After all, Europe - and especially Great Britain - could not survive long without Confederate cotton. Hunley knew better. He would have known there were efforts being made even then to start cotton crops in other parts of the world (most notably Egypt), and the longer a blockaded Confederacy had to sit on its cotton, the more likely those other crops were to be planted.

Hunley understood the high technology of that era - the machines needed to process the cotton, the ships and railways needed to transport it, and the science used to continually improve it. He also understood that for all the bluster and

The Long Patrol

oratory, the CSA's chances were not good - little usable manufacturing capability, no fleet, and few natural resources save for manpower, and even that wouldn't last forever. Hunley knew instinctively that for the Confederacy to survive the sea-lanes to Europe and the Indies had to stay open at all costs, and that there was little if any chance for the Confederacy to survive long enough to build or acquire a fleet that could challenge the USN.

Just after the outbreak of war, Hunley took command of the schooner *Adela* and headed for the Yucatan to meet two other vessels that were supposed to be bringing weapons and ammunition to the Confederacy. Hunley never did find the two ships, but he got a crash course in the difficulty of running a blockade, even one as porous as the USN was running at that point. Whatever else one might say about Horace Hunley, one can't fault his devotion to the Cause - Hunley refused any payment for the trip, since, as he pointed out, he was unsuccessful. He later sailed to Cuba to both purchase whatever weapons and resources he could and arrange for more to follow through the same route.

Cuba, Mexico, and the Indies would be busy, thriving ports of call for blockade-runners until the last Confederate port fell. The friction between the British Empire and the US during the Civil War is well known, so that explains why Jamaica and the Bahamas were so popular with the runners. Spain probably had no serious disagreements with the US, but their doddering Empire was unable to keep an eye on everything and it seems quite reasonable that more than a few Cuban *alcaldes* and *corregidors* decided to take advantage of the situation and make it easy for the blockade to be broken through their ports. After a bit of profit sharing, of course.

Hunley had made it quite easily in and out of New Orleans – 'ease' being a relative term. He'd gotten to Havana, closed a few deals, and made it back with a few more precious rifles and other supplies, and even won a commendation for his actions.[9] The blockade had been far from tight then, but Hunley knew that

---
[9] Ragan, pg. 18.

CSS H.L. Hunley

wouldn't last forever even with the converted tubs that were being pressed into service.

The answer was simple and direct – the blockade must be broken. But how? And how could it be done while still turning a reasonable profit? That should not be held against him, far from it. After all, he was a businessman in an era where profit was revered – even worshipped, and the great robber barons were making their reputations. Patriotism was fine, and if there was a bit of money to be made, so much the better.

We don't know if Hunley ever read Reverend Smith's letter, but is it very likely that he did. The letter appeared in newspapers throughout the South, and Hunley was a well-read man. The timing would have been just about right – Hunley returns from Cuba, deeply concerned about how the blockade was going to even be challenged much less broken, and there in the pages of the New Orleans *Picayune*, was Reverend Smith's commentary on submerged combat.

Knowing Hunley, he would have considered such a revelation a demonstration of the Lord God's power to save his chosen – and without question in the summer of 1861, Horace Hunley considered the Confederate States of America the chosen. Here indeed was a possibility – an extreme one, to be sure, but in an era where nothing was considered impossible to men who would simply try hard enough, it was sufficient. And Horace Hunley knew – he *knew*, with every fiber of his being, that he could pull it off.

To paraphrase secessionist William Yancey as he introduced Jefferson Davis, the man and the hour had met.

II: "...WHOSE ARM DOTH REACH THE OCEAN FLOOR..."

Up until the early twentieth century, summers in New Orleans had a tendency to be dangerous at best. Yellow fever and malaria were still regular and efficient killers there - some years taking as much as sixty percent of those infected and killing as much as fifteen percent of the overall population. Close to thirty thousand people would die of the Fever from the first reliably recorded epidemic in 1817 until the last one in official one in 1905, and those are just the recorded deaths in and around the city proper.[10] There was also the occasional *Ouragan*, hammering the city at regular intervals, though another century and a half would pass before a long week of horror called Katrina in 2005. And of course, there was the River, which was known to burst through the towering levees and scour entire neighborhoods from the face of the earth with water from storms that no resident ever even saw.

Wondering whether or not the weather or some unseeable organism or a twelve-foot wall of stinking water was going to kill you added a certain excitement to things. But as any native of the Crescent City will tell you, they are an easygoing and adaptable lot and death becomes part of life, *c'est la vie*. The summer of 1861 though was even more dangerous than usual, for New Orleans was a vital strategic port for a nation trying to survive – and this time, a target in the most literal sense.

A look at the geography is probably in order. The Stars and Bars flew from Cape Hatteras in the east to West Texas, and the Mississippi River was simultaneously a highway, a defensive line, and a highly vulnerable jugular vein for the Confederacy. The problem was that the Confederacy's armed forces – especially the Navy - didn't have the numbers or mobility to defend it properly. The Confederacy's leaders knew that once

---

[10] George Augustin, *History Of Yellow Fever*, Searcy and Pfaff, New Orleans, 1909.

CSS H.L. Hunley

Old Man River was severed – or worse yet, in Federal hands from one end to the next – then it could be isolated, strangled, and eventually overrun.

The Federals simply could not be stopped from the northern end of the river – there would never be enough manpower or firepower or sufficient means to get them there and keep the Confederate forces supplied. The CSA and CSN riverine forces had the ability to challenge and defeat the Federals at local spots along the river, but they would have to rely to a great extent on luck and the incompetence of early Federal commanders, something else that wouldn't last forever. What had to be done was to keep New Orleans as a viable port and military base. As long as it was, the Mississippi would at least be contested and there would still be a link – however tenuous – to the Trans-Mississippi states.

Trouble was that the Federals knew it too. So much so, in fact, that they made the Mississippi River a vital part of their ultimate plan to kill the Confederate States of America. Given what they had to work with at the time, US Army commander General Winfield Scott – once the strong, virile victor of the Mexican War, but now an old, quivering caricature of a warrior – realized that the calls of "On To Richmond!" were not only hollow, but self-defeating as well. From his perspective, he felt that such a strategy would end in a war of conquest that would leave a victorious Union with 'fifteen devastated provinces.'[11]

What General Scott had in mind instead was what he called the 'Anaconda Plan.' Rather than charge headlong into the Rebel States, he would surround it with a sea blockade and a riverine fleet on the Mississippi that would slowly squeeze the life out of the Confederacy, like the snake it was named after. This plan would eventually become the default Federal strategy, but it failed to satisfy the louder voices in the Union who were calling for an immediate assault on Richmond and Scott – who was a Virginian by birth and whose Unionist sentiments were therefore somewhat suspect – was carefully eased out of the

---

[11] James M. McPherson, *Battle Cry Of Freedom*, Oxford Univ. Press, New York, 1988, pg 333.

decision-making process, though the Anaconda Plan would survive its creator and lay the groundwork for the eventual defeat of the Confederacy.

A blockade needs bases, and the Union was determined to get them. Originally, the blockade was run from Union bases in New York, Philadelphia, and Baltimore – the old Gosport Yards, now the Norfolk Naval Base, had fallen into Confederate hands and would stay that way until April of 1862. In addition, the Union bases at the head of the Mississippi were an unshakeable anchor for the forces that would eventually batter their way downstream. But to truly make the blockade work New Orleans had to be taken off the board, and the Federals were working on that from the very beginning of the war. It would be a steady process though for the Federals had the men, the ships, and the materiel. They could afford to take their time.

Horace Hunley was working too, and he had far less time than the Federals did. Although the Confederate press routinely portrayed the war news in glowing terms of success and unending victory, Hunley was no fool. He had sufficient intelligence, experience, and – perhaps most importantly – connections to know what was happening out there beyond the marshes of the Delta, and he knew it wasn't good. The blockade was getting just a little tighter each day, just a bit more secure, and eventually one day the United States Navy would come through that line followed by the United States Army and head for New Orleans itself. And when that day arrived, it was not at all likely that the returning Federal authorities would be understanding about the city's activities since secession in general…or specifically those of one Horace Lawson Hunley, Esq.

Now, it should be noted that Hunley does not seem to have ever done anything more sinister than the average Southron - he wasn't one of the leaders of the secession movement, nor was he one of the firebrands who urged others on - and there is no reason to suppose that the Federal Government, in its infinite wisdom and vengeance, had any particular reason to go hunting for him. But men who challenge Governments sometimes believe that they have angered men with the power to hurt them,

CSS H.L. Hunley

and Hunley now added - had to add - sheer self preservation to his other concerns.

Hunley knew what he had in mind – a submarine boat that could go out to sea under her own power, sink Union warships, and come back. What he didn't know was how in the world to build one. The answer to that would seem simple enough – find a shipyard and have it built. New Orleans had shipyards, and excellent ones at that. They however were solidly committed to supporting the CSN's war effort and every man, piece of timber, and sheet of iron was spoken for. That left him with one option – building it himself. But even for a man of Horace Hunley's talents, that was going to be a stretch. He was going to need, at the very least, a decently sized machine shop and the space to put something together. He was also going to need someone who knew something more about ships and sailing than just the proper terms for the left and right. Fortunately, he found one of each in short order.

James McClintock
Source: U.S. Navy

We don't exactly know how James McClintock and his partner, Baxter Watson, met Horace Hunley. McClintock and Watson had already made something of a name for themselves in supporting the Confederate war effort by providing desperately needed parts for the relative handful of steam railroad engines that were providing the only strategic mobility and supply the CSA had at the time. They were also inventors, having come up with a machine that could make hundreds of bullets at one sweep.[12] McClintock was also the holder of a unique distinction – he was the youngest licensed riverboat captain on the Mississippi River. The popular image of the

---

[12] www.thehunley.com/Coatsesssay.htm

majestic sternwheelers gliding down the river belies the fact that they were often difficult and temperamental machines to handle under the best of conditions, and when throwing in wind and the unusual currents and eddies that were prevalent in the old Mississippi (after more than a century of work by the Corps Of Engineers, even the present day Mississippi is still a tough, unforgiving maritime environment that regularly claims unwary captains and their vessels), and for anyone under their late forties to show a mastery of the river was extraordinary.

In other words, James McClintock knew what he was doing afloat, had experience in dealing with the high technology of that era, and had access to a facility that could work with that technology. For Horace Hunley, it must have seemed as if the Gods were smiling down upon him and his idea. Sadly, we don't know much at all about Baxter Watson, other than that he was McClintock's partner. It does seem reasonable however to suggest that he was probably as skilled in matters technical and mechanical as was Captain McClintock.

"...The solid Leeds Foundry in New Orleans..." The Pioneer was designed and built here in the weeks before the Federal assault on New Orleans. (Photo courtesy of www.neworleansonline.com)

Of course, any project like this would take time and money. We already know they were short on time, but money was a different matter, at least for the time being. Hunley's fortune under the Union was dwindling under the Confederacy, but it was still an impressive amount of money by any standard in use in 1861. With that in mind, it should have been fairly easy for Hunley to hand over $400.00 CSD to McClintock and Watson for them to begin work on what they intended to be the world's first combat submarine.

CSS H.L. Hunley

The sub's birthplace may have been the solid old Leeds Foundry in New Orleans – founded thirty-seven years before at 923 Tchoupitoulas Street and still there today as a National Historical Landmark. There is some argument over this - the US Naval Historical Center mentions that McClintock said afterwards that she was built at his shops on Front Levee Street. Wherever they did it, we can safely assume that the three men sat down and carefully worked out what they wanted to do and how they were going to do it, based on the limited resources they had. When they started is also unknown to us, but it had to have been sometime by the early fall of 1861.

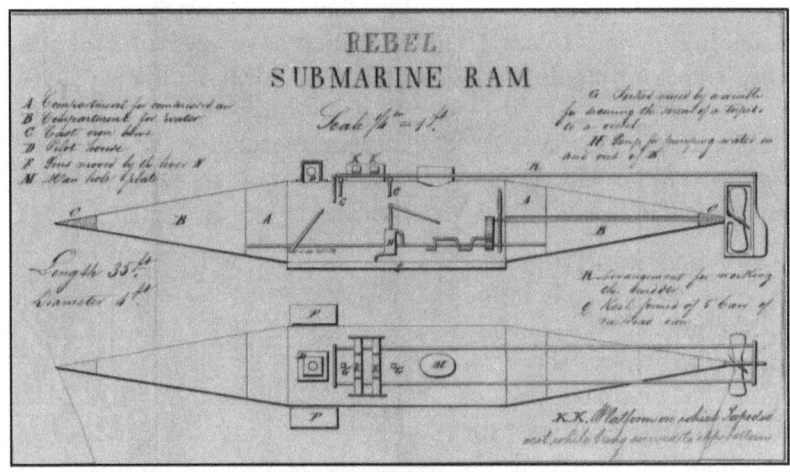

"The first – and thus far, the only – submarine ever to be commissioned as a privateer…" Plan view of Hunley and McClintock's first submarine, the CSS Pioneer. (Source: National Archives)

This is what we know of the boat they built that warm, muggy fall: It was between twenty and thirty feet long, built from sheets of cast iron that had been cut from old boilers (a theme we shall return to) and bolted to an iron framework. The boat was of mostly cylindrical cross-section and tapered to sharp points at either end. There was a small conning tower topside and small glass ports – charmingly called "deadlights" – along her flanks. She was powered by one man, turning a crank that was attached directly to her propeller. She had diving planes and a rudimentary ballast system. Her punch – called a torpedo, but what we would now call a mine – was a canister of black powder with a clockwork fuse, carried on her back.

The idea was that the boat would slip under its target preferably while it was stationary. The crew would then drill a hole into the bottom of the ship, attach the torpedo to it, set the timer, and sail away. A few minutes later there would be a huge explosion, the hated Federal ship would sink without apparent explanation, and the blockading fleet would flee in stark terror. At least that was the plan.

"...And make a fortune - if they could make it through the blockade..." The remains of the blockade-runner Ruby, ashore on Folly Island, SC, after trying to evade the Federal blockade. (Source: Library Of Congress)

The boat was ready sometime before March of 1862, but there does not seem to be a record of her trials. This isn't terribly surprising; the boat – by now christened *Pioneer* – would have been considered a secret weapon of the first water, and there wouldn't have been a lot of people authorized to see her tested. McClintock stated after the war that there were several test runs made against moored target ships that ended in their destruction. Now, ships suddenly exploding for no apparent reason in normally peaceful Lake Pontchartrain should have drawn someone's attention, but never mind.

In any event, there appears to have been general satisfaction at the performance of the sub, and some Confederate military personnel had seen *Pioneer* in action by this point. The poor soul who had to crank her through Pontchartrain would certainly

CSS H.L. Hunley

have been getting a workout, and apparently no one had considered the effect of an iron hull on a compass,[13] but once allowances had been made for that they were doing quite well.

It was at this point that McClintock, Watson, and Hunley decided that *Pioneer* was ready for the big leagues and formed a syndicate to send her to sea as a privateer. It's not entirely clear exactly what they had in mind – a privateer, after all, must be able to capture ships and bring back their cargo. Under the best of conditions, *Pioneer* would only have been capable of a couple of knots per hour, and under no circumstances would she have been able to capture a ship at sea.

What seems likely is that Hunley and his crew wanted a legal fig leaf to cover themselves in the event of capture - although privateering was, as previously mentioned, illegal to the people they were going out after, attempting to destroy enemy merchant ships as civilians was even worse. Civilians who do that are called 'pirates' and were subject to summary execution upon capture. But if Hunley and his crew could show a letter of marque, they could claim to be entirely legal under Confederate law, and since the US Constitution still authorized letters of marque they might be able to confuse the issue long enough to survive captivity. That was, at best, a slender hope, especially if the *Pioneer's* crew was captured after sinking a Federal ship with heavy casualties. But this was an era of legal niceties, so one can be sure that had *Pioneer* ever ventured out to sea, that letter of marque would have been tucked securely in someone's vest pocket.

In any event, quite a group came together as backers – Horace Hunley, of course, along with McClintock and Watson, plus Hunley's brother in law R. F. Barrow, a J.K. Scott, who served as *Pioneer's* skipper during her trials, and one Henry Leovy, a college classmate of Hunley's. They put up a $5,000.00 CSD bond and submitted a request for a letter of marque to the CSN at Richmond. There, F.H. Hatch, a Confederate official and old friend who had commended Hunley for his trip to Havana, authorized the letter of marque

---

[13] Ragan, pg. 20.

on or about April 1st, 1862. CS Submarine *Pioneer* therefore became the first – and thus far, the only – submarine ever to be commissioned as a privateer.

Just three weeks later, Hell came to visit New Orleans.

It is said that when the Union fractured, Southern naval officers tried to convince Tennessee-born Flag Officer David Farragut to join them. His answer was as succinct as it was foreboding: "Mind what I tell you: you fellows will catch the devil before you get through with this business."[14] Well, the Devil Himself brought the USN over the sandbars at the mouth of the Mississippi in early April, confounding the Confederates who thought Farragut couldn't bring a sufficiently powerful force into the river to threaten New Orleans. As it turned out, the warships alone were the least of the Big Easy's worries.

Admiral David Farragut. Source: U.S. Navy

Farragut had specially built mortar schooners brought up to hammer the primary coast defenses of New Orleans, Forts St. Philip and Jackson. Built along the same lines as Fort Sumter (though smaller) they were intended to duel with attacking ships trying to force their way upriver. Nobody had ever thought of an enemy who might stand off outside the range of their big guns -- 10 inch Rodman cannons at Jackson with a range of about two miles - and simply turn the massive fortifications into mounds of shattered brick.

Farragut's plan was for the mortars to take the forts completely out, but after not quite a week of continuous pounding that seemed to have little real effect (other than to permanently deafen many of the mortar crews) he decided to do it the old fashioned way. At two o'clock in the morning on

---

[14] McPherson, pg. 419.

CSS H.L. Hunley

April 24th, Farragut ran the forts. It must have been a sight – nearly one hundred large-caliber guns in the forts, two hundred in the fleet, all roaring away as fast as their crews could fire them. The Confederate ironclad CSS *Louisiana*, unable to steam but with all her guns operational, did the best she could while moored to a dock. A fleet of Confederate fire ships, loaded with gunpowder, explosives, pitch, and anything else that would burn was set alight and sent downriver directly towards the oncoming fleet.

A Lithograph showing Farragut's fleet "Running the Guns". The chaos and confusion of a night action on the Mississippi is captured perfectly. Source: U.S. Naval Historical Center.

And by 3:30 AM it was all over. Farragut's fleet had been hit hard with almost two hundred casualties, but they were past the forts and safe. *Louisiana's* crew, who had shown incredible bravery in fighting outnumbered and outgunned while their ship was immobile, did the only thing they could and scuttled her a few hours later. Her eventual whereabouts became a mystery of their own until author/shipwreck finder

The Long Patrol

Dr. Clive Cussler found her in 1981.[15] The garrisons at St. Philip and Jackson mutinied and fled before dawn, taking the militia with them. Fifteen thousand Federal soldiers, led by the rotund and cross-eyed Benjamin Butler, then landed from the Gulf with no resistance whatsoever and headed north.

New Orleans, only seventy miles away, lay utterly defenseless. It took less than a day for Farragut to make his way upstream, where the river batteries at least tried to stop him, but 11" guns on the US ships firing at point-blank range were able to silence them as easily as a man swats a fly. By noon on the 25th, David Farragut was gazing down upon the Crescent City. It was another four days before the Marines went ashore, but for all practical purposes the fight was over. The most important port of the western Confederacy and the southern anchor of the Mississippi were gone.

As the guns slowly boomed their way upriver, Hunley, McClintock, and Watson were quietly getting ready to evacuate New Orleans. Even going by the most optimistic of reports New Orleans' hours were numbered, as were those of the *Pioneer*. She was a weapon of amazing potential, even as limited and basic as she was, and there was never any question that she could be allowed to fall into Federal hands.

It's unlikely that *Pioneer* ever made any actual patrols before the fall of New Orleans. She almost certainly would have had to have been transported quite some distance in order to have even had a realistic crack at anything, and it should be pointed out that there are no records of any testing in open ocean. She would have handled much, much differently there than in the relatively sheltered waters of Lake Pontchartrain – if she could have been handled at all, but in the event there was no way to ever find out.

Hunley, McClintock and Watson certainly would have wanted to get their creation out of town, but *Pioneer* could only have been moved by train – she was far too heavy to have been moved by any other means, and every train in the area was being

---

[15] Clive Cussler, *The Sea Hunters II*, Berkley Books, New York, 2002, pg. 106.

loaded to overflowing with troops, materiel, and People Too Important To Be Captured. The three designers could get out easily enough, their notes, plans and concepts with them. But their creation could not escape, any more than the statue of Andrew Jackson down in the French Quarter could gallop away. Taking her apart and hiding her was no option either, for it had taken months just to put her together. That left only one choice, but it would be a fitting end for a warship - even one as humble as the *Pioneer*.

There was a four-day window between Farragut's arrival offshore on the 25th and the Marines finally raising the flag over City Hall on the 29th, and from all accounts it was a fever-dream experience. Much of the time was occupied by drawn-out negotiations between the city fathers and Admiral Farragut, negotiations that more resembled an Abbott and Costello routine than anything else. In the meantime, New Orleans was about to fall to an enemy force, and the residents were believing every word they'd been told about what the devil Northerners would do once they arrived – burn the town, ravage the women, etc. With that in mind, the citizens were burning tens of thousands of bales of cotton, tons of gunpowder, and other supplies in a mad effort to keep them out of Federal hands.

"Trouble was, it wasn't the Pioneer..." The submarine pulled from St. John's Bayou after the war, possibly a boat ordered from Tredegar Iron Works by the Confederate Government. (Photo courtesy of the Louisiana State Museum)

With the United States Navy in sight, the sky darkened by smoke, and the populace running about in fear, Hunley, McClintock and Watson gathered at the old New Basin Canal, which used to run from Lake Pontchartrain down to a turning basin near the present Union Passenger Terminal. There,

McClintock climbed aboard *Pioneer* one last time and opened her ballast valves. Quickly scrambling out, McClintock gave her a push and the little submarine bobbed away from the shore, rocking gently as she drifted out into the canal. Each time she rolled, the movement was a little slower, a bit more sluggish, until after a few minutes *Pioneer* sank gently beneath the brown water, a few stray bubbles marking what her builders hoped would be her final resting place. With that, they headed for the train station and caught one of the last trains out, destination Mobile, Alabama. Mobile had everything New Orleans had – modest shipyards, fabrication and machine shop facilities, manpower, and at least for a little while, safety.

There is a postscript to that somber ceremony on the bank of a Louisiana canal one dark April day in 1862. In 1878, workers dredging out St. John's Bayou discovered a submarine boat and pulled it ashore. Although what happened next is unclear, the boat raised from the river eventually ended up at a Confederate veterans' home, then the Louisiana State Museum where it remains today. The *Pioneer*, it seemed, had found her niche in history. Trouble was, it wasn't the *Pioneer*. The *Pioneer*, it seems, had been raised from her final berth in 1868 and sold for scrap a few days later.[16] Mark Ragan, in the process of researching his superb book on the *Hunley*, finally settled that question once and for all in the late 1990s.[17] So the question now is - what submarine is this?

There are several possibilities, outlined in a fascinating article by Brian Hicks of the Charleston *Post and Courier*.[18] Hicks believes that it was built by two men named John Nesmyth and John Roy, who were even experimenting with cannon and mortars mounted on the little boat. What is even more interesting is that this boat – and perhaps as many as six other boats being assembled in Louisiana at the time – were being built in ignorance of each other. Imagine what the infant

---

[16] Angus Konstam, *Confederate Submarines and Torpedo Vessels 1861-65*, pg 46. Osprey Publishing, Oxford, UK, 2004
[17] Ragan, pg. 25.
[18] http://www.ussvi.org/mem/mstrysub.htm

CSS H.L. Hunley

Confederate Navy would have been able to do if these efforts had been pooled. Another one of these boats was called *Saint Patrick,* and although there is little solid information on her, by all accounts she was even more advanced than the Hunley/McClintock/Watson boats – but for reasons unknown only saw action once, nearly a year after *Hunley's* loss. And there is yet another possibility, which we will return to shortly.

Mobile in the summer of 1862 was a grim place, with nowhere near the *joie de vivre* that New Orleans had at least tried to maintain to the end. The units that had retreated out of New Orleans had arrived demoralized, hungry, and in some cases almost without weapons. Refugees – many of whom had nothing to fear had they stayed in New Orleans, but a few with guilty consciences who feared retribution from the Federal Government - had also flooded in as well. CSA General Dabney Maury, nephew of the CSN's Matthew, was doing his utmost to hold things together there as the city's commander, though even he admitted it would be touch and go. A tough, courageous officer whose short stature and ferocious courage earned him the affectionate but slightly unusual nickname of 'Puss in Boots', Maury was faced with a daunting task that he could accomplish for a while, but no more than that. There was no question that Mobile – with the finest natural harbor on the Gulf of Mexico – was in the sights of Federal forces, and their arrival was not far off. The only question was when.

One of the better mechanical facilities in Mobile that summer was the Park and Lyons Machine Shop at 250 North Water Street, just a thousand feet or so from the Mobile River and now the site of the Alabama State Port Authority. The shop, a solid, straightforward brick building that survived into the 1960s, was within easy reach of the shipbuilding and fabrication facilities that lined the riverfront, and they were doing the nation's work – to be precise, rifling musket barrels. It was here that Messrs. Hunley, McClintock, and Watson showed up one day, certainly in the late summer or perhaps even early fall of 1862, carrying rolls of blueprints and reams of notes. General Maury had specifically sent them there with a unique set of orders for the proprietors and the CSA personnel who were attached to the shop – drop what you're doing and assist these three men.

The Park and Lyons Machine Shop. Source: Library of Congress

Maury was nothing if not willing and eager to embrace new thinking and new technology. Some of it was out of necessity – he only had about 10,000 troops to defend the Mobile area, and that was perhaps a tenth of what he needed. Afloat, there was almost nothing at all. Some well-built and powerful ironclads were there, but they were badly outnumbered and outgunned by what the USN would be able to bring down on them. There was a bit of interesting technology awaiting the US Navy out in the harbor – a torpedo field (what we would call today a moored minefield) that would hopefully attrite any attacking force to the point where it could be defeated in detail by the ironclads and the harbor defenses, or better yet, deterred completely. But the torpedoes were a sometimes temperamental, and mostly unproven technology, so Maury realized he needed a couple more aces up his sleeve.

And that was where the other part of his willingness to try something new came in. His Uncle Matthew was already a legend for his work on oceanography as well as charts and meteorological work.[19] But when he left the USN to join the Confederacy, he was sent to England to secure ships and harbor defense systems, where he did some experimental work with torpedoes. This information was passed back to his nephew the General, who – to use a modern phrase – ran with it, and it also got him thinking about other new technologies that could be used to defend Mobile.

There would have been many a war council with the Confederate commanders, always trying to find something that they could use. Eventually, someone would have mentioned those three men and their submarine privateer, and at that point General Maury's attention would have been assured. Once again, the story is vague here – we don't know how the three designers met General Maury, we don't know how they made their pitch for the scarce resources and manpower under Maury's command (though one would love to have heard it), and we don't know who else was involved in the decision. Whatever took place though impressed General Maury enough that he sent them to an important engineering facility with orders to take it over and get to work.

A fourth member of the story joins us here – Lieutenant William Alexander, Confederate States Army and late of Co. 'A', 21st Alabama Infantry Regiment. The 21st Alabama had been badly hurt at Shiloh, sufficiently so that rebuilding it as a coherent unit may have been out of the question. It also was raised in the Mobile area, so sending what was left back home was a reasonable idea. Alexander, originally born in England, was an engineer by training who had immigrated to Mobile in 1859. When the 21st Alabama got to Mobile, Alexander was assigned to Park and Lyons as the CSA's representative on the barrel-rifling work, and that was where he was serving his adopted country when the fleeing inventors arrived. There was another member of the 21st Alabama there as well – George Dixon, a former enlisted man who had quickly risen to the rank of Lieutenant himself, and was still recovering from a serious leg

---

[19] http://xroads.virginia.edu/~UG97/monument/maurybio.html

injury at Shiloh. We shall hear more of Lieutenant Dixon later.

Within a few weeks, McClintock, Hunley and Co. had laid down the blueprints for their second boat, the next step in their efforts to sweep the seas of the hated Federals. We know even less about this second boat than we do about little *Pioneer* or the *Hunley* herself. She was longer than *Pioneer*, though not by much. Her primary hull is adapted from a boiler – after all, it's already the right shape and length, so the builders made a virtue out of necessity. To look at the few extant drawings – done some time after the war – she was much more streamlined, a definite change from the Jules Verne appearance of the *Pioneer*. But to look at her now, we can see so very, very clearly the first outlines of the boat we came to know as the *H.L. Hunley*.

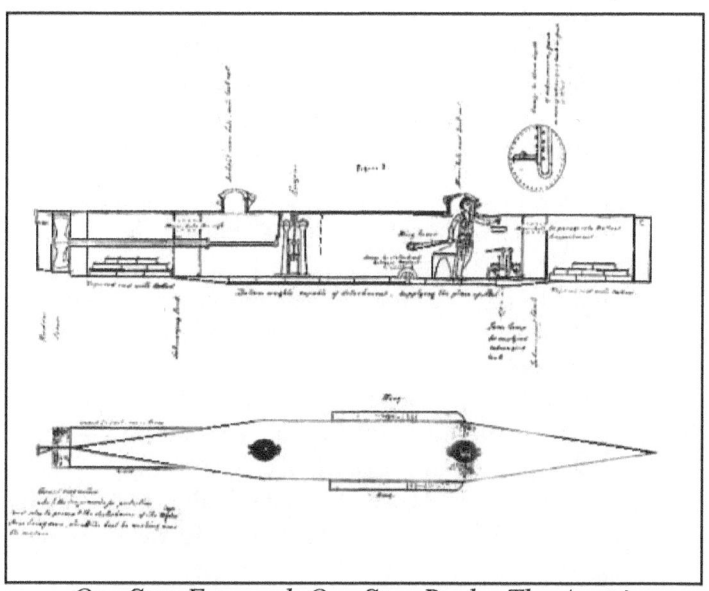

One Step Forward, One Step Back: The American Diver/Pioneer II, in a drawing done by James McClintock for the Royal Navy. American Diver/Pioneer II still rests somewhere in Mobile Bay, victim of an unexplained leak while being towed to Fort Morgan, AL, to attack the Federal Blockade. (Photo courtesy of the British Admiralty, Records Division)

There is the long, tapered hull with their streamlined casings and the two low conning towers with small glass ports. There are the two long, narrow diving planes amidships, the propeller ahead of a circular iron ring, and the control rods that lead to a rudder behind it. Iron weights are bolted to the bottom of the hull, and there is a primitive depth gauge like that which will be found inside *Hunley*. She is streamlined, solidly built, and carefully laid out, the end product of the long series of test runs done with *Pioneer*.

In short, she is a cutting-edge weapon, the best that Confederate naval technology, engineering skill, and a technologically interested and supportive military leadership can provide at that time and place. We don't know her exact name – she is referred to in various texts and sources as *Pioneer II*, *American Diver*, or even *Pioneer II/American Diver* (which, truth be told, is still better than the name given to her Federal counterpart under construction in New Jersey – *Intelligent Whale*[20]). No matter what her formal name was, however, what they called her could not have been complimentary. The Mobile boat was a difficult, frustrating failure from the start.

On the face of it, the Mobile boat should have been a step forward from *Pioneer*, but she wasn't. First was the matter of her propulsion. She appears to have been somewhat larger than *Pioneer*, and therefore heavier – but the same human-powered propulsion system was still in use, albeit with four men on the crank. Unfortunately, it wasn't enough to get any reasonable speed up, and it became clear very early on that she would have to be towed anywhere more than a mile or so from her moorings – a problem that would eventually doom her. There are tantalizing suggestions, however, that Hunley and McClintock had originally designed the Mobile boat to have a high-tech solution to her propulsion – an electric motor.

Some time after the war, McClintock told the British Admiralty that considerable time and effort went into the design of an electro-magnetic motor that would, had it worked, have literally taken the Mobile boat fifty years into the future. Able to move at even four or five knots on her own, against currents and wind, she would have easily been able to outmaneuver and in

---

[20] See Appendix 2.

The Long Patrol

some cases even catch some of the slower blockaders, and do it almost unseen and with the ability to dive safely out of reach.

In short, the Mobile boat would have been – at least in terms of mobility – surprisingly close to the boats that were in service at the beginning of the twentieth century. Her time submerged still would have been limited – the Mobile boat does not appear to have had snorkels, so her crew would have been running on whatever air was trapped in the boat. But without the need for five men, four of whom would have been turning the propeller, the Mobile boat might have been capable of spending more than three hours underwater with only two men aboard.

But for all its possibilities, the electric motor planned for the Mobile boat would remain just out of reach. The US Navy says that "much time and money was spent", and that Hunley himself stated that he provided "the entire means" for building the Mobile boat, [21] and it appears a great deal of it went to the little engine that couldn't. (This, by the way, would tend to support Mark Ragan's findings that the Confederate States of America spent very little, if any, actual funds on construction of the three boats. [22] There was a reason for that, however, that shall be examined later.)

The technology was there, the technical knowledge was there, but the ability to put it all together was not. Perhaps in New York or Boston or Baltimore, but not in tightly blockaded Mobile waiting for the crack of doom. (One wonders though what the result would have been had a single electric motor and batteries somehow made it through the blockade from Great Britain.) With the electric motor out of consideration, that left simple manpower and that simply would not be enough.

Secondly, the Mobile boat apparently did not steer well either. Having been intended to move under steady motive power, once she was refitted with a hand crank she would have been sluggish at best and nearly uncontrollable in anything more than a glass-calm sea.

---

[21] See Appendix 3.
[22] Ragan, pg. 27

CSS H.L. Hunley

Third, her weapon of choice left something to be desired. *Pioneer* had a small clockwork mine that was carried on her back, to be attached to the hull of her target using an auger and screws. While this might very well have worked in a technical sense, practically it left a great deal to be desired.

First, it required a motionless target. That was probably a detail that could not be worked around, as even an electrically powered boat would not have been able to get up enough speed to execute an intercept-type attack on a moving vessel. Secondly – and perhaps more problematically – it required the crew of said motionless ship to be completely unaware of the sounds and motions of someone trying to bore a hole in the hull of their vessel.

McClintock, Watson and Alexander seem to have realized this as well, and came up with another idea: they would trail a torpedo on a line some distance behind the boat. There are differing guesses as to just how far, but one hundred and fifty to two hundred feet appears likely. They would dive just before reaching the target and pass under it, while the torpedo, still riding on the surface, would be pulled up against the target's hull. Once the torpedo line went taut, a crewmember on the boat would pull a firing lanyard and the torpedo would detonate. Given the inherent limitations of the boat, it was a pretty elegant compromise – but as it turned out, an extremely risky one.

Initial tests indicated that without question the idea would work – a few more derelict barges gave their lives in the Mobile River, seemingly struck by lightning from a clear blue sky. However, as is so often the case with a new idea, it turned out to have some practical limitations.

Once the boat started moving further offshore, it was discovered that the torpedo had a disconcerting habit of refusing to obediently follow the boat. Wave action and wind were often enough to send the torpedo veering wildly off to one side, and in some cases would even cause the torpedo to overtake the boat itself.

Finally, there was one last flaw that would lead directly to the eventual loss of the Mobile boat. In order for her to use the torpedo in an attack, she would have to dive under her target. That was easy enough against a shallow draft barge, but when it came to deeper draft ships, it was discovered that there was nowhere in Mobile Bay that was deep enough for the sub to dive deeply enough safely. In short, the Mobile boat drew too much water to execute a useful attack in the very waters she was intended to defend. Given the enthusiasm of her designers and builders, it seems reasonable that they would have worked hard to overcome these obstacles to the greatest extent possible. However, there was now someone keeping an eye on their activities and expenditures – and it was not an entirely approving one.

In August 1862 Admiral Franklin Buchanan, CSN, had been assigned to Mobile as commander of CSN forces on Mobile Bay. Buchanan was an Old Salt in the best tradition – he had gone to sea just after the War of 1812 and had been serving ever since. His reputation for skill and integrity was such that in 1845 he was named the first Superintendent of the US Naval Academy. Buchanan was certainly no stranger to technological innovation – he had been the first CO of CSS *Virginia*, the ironclad built upon the charred bones of USS *Merrimack* after she was dragged from her shallow grave at Gosport Navy Yard.

Admiral Franklin Buchanan, CSN. Source: Naval Historical Center

In Buchanan's hands, the *Virginia* had inflicted more damage in one day on the US Navy than any other enemy. *Virginia* sank the frigate *Congress* and the sloop *Cumberland* and decimating their crews. Bringing the unwieldy ironclad around, Buchanan then ran the frigate *Minnesota* aground before Federal gunfire wounded him so severely that the Executive Officer broke off the fight to evacuate him ashore. That *Virginia* was unable to finish its work the next day was due solely to the

CSS H.L. Hunley

arrival of USS *Monitor*, which arrived the next day with timing that would have done credit to a stage hero. Less heavily armed than *Virginia* but faster and more maneuverable and equipped with a revolutionary rotating turret, *Monitor* managed to end the fight in a sort of Mexican standoff, and *Virginia* had to pull back to the safety of her lair at Gosport. There, on May 11th, *Virginia* was scuttled by her crew off Craney Island to avoid capture by approaching Federal forces.

McClintock and Buchanan represented different eras and philosophy of warfare. It was hardly surprising they had a failure to communicate. Source Naval Historical Center.

Virginia's fate could in no way be laid to Frank Buchanan. Although he was not aboard during her climactic fight with the *Monitor,* her crew handled her superbly. But *Virginia's* steam engines, none too reliable to begin with, were almost useless after the fight. Even if they had been in better shape, there was nowhere for her to run – she drew far too much water to go upriver to Richmond, and she could not survive in the open ocean. So, after his ship was blown to bits, Buchanan was transferred to Mobile with a promotion to Admiral and a mandate to organize the naval defense there. The submarine projects would have come directly under his purview.

And quite frankly, he wasn't all that keen on it. It wasn't that Buchanan had a problem with new technology, in and of itself - he'd been a little skeptical when he took command of *Virginia* back in Hampton Roads, but he'd realized very quickly that he had something that was going to change history. But this was…well, impractical. The most powerful punch the boat could throw wouldn't equal the power of two or three cannon, and Buchanan knew from bitter experience that it took a lot more than that to even cripple a well-built merchantman, much less a warship.

But above and beyond that, Buchanan came from a world where honor and reputation was everything. As a boy he would have known men who sailed with John Paul Jones, or taken the *Guerriere* with Isaac Hull, men who had stood unflinchingly on a quarterdeck in a hail of cannon fire while musket balls whirred by and literally gone face to face with an enemy. You couldn't do that with a submarine – indeed, it was impossible. Buchanan certainly understood the potential of the submarine, but he considered it somewhat ungentlemanly . . . unfair. However, he had his orders, and Frank Buchanan was a man who followed orders. He provided all the assistance he possibly could to the builders at Park and Lyons, but he was never one to hide his opinions. Hunley, McClintock and Co. would have known quite well where the old Admiral stood.

Buchanan's feelings would have been both a rebuke and a challenge to the builders – a rebuke insofar as a great amount of time, effort, and money had been put into it, and so far there was damned little promise to the narrow iron boat. On the other hand, the challenge was clear: prove to Buchanan that it could work...it *would* work. It would have to work.

That seems to be the best explanation for what happened next. During the late fall of 1862 and the early winter of 1863, the Mobile boat was being put through her paces and her shortcomings were being made painfully clear. McClintock and Watson apparently made the decision at this point – early February - to send her out on her first patrol. A Confederate deserter claims that she went out on or about Valentine's Day, but this seems more like a garbled mix of witnessing a test run and hearing some fairly solid reports of what was planned.[23] What appears more likely is that sometime after February 14th, she was prepared to head out to sea. Due to her inability to make any kind of speed and her requirement for deeper water, she was

---

[23] Ragan, pg 33.

to be towed out to Fort Morgan, the massive stone keep that guarded the approaches to Mobile Bay. There, a crew would board her and take her out into deep water to attack the blockaders.

This was a mission the Mobile boat never should have undertaken. She was far too slow, she was difficult at best to control, her torpedo stood a good chance of striking her instead of her target, and no one was entirely sure of the water depth where she would make her run. This was needlessly hazardous at best, and suicidal at worst. But McClintock, Watson, and Alexander seem to have been determined to get her to sea and in action, and she went.

The weather was less than ideal, but apparently within whatever standards had been set for her operation. The small launch assigned to tow her nearly forty miles from the Mobile waterfront out to star-shaped Fort Morgan got underway smartly and turned south into the Mobile River and downstream to the bay. They would have passed within sight of where the battleship USS *Alabama* sits today, maintaining a silent watch over the city that adopted her a century later. Once they got down to the open waters of the bay, however, things started to go terribly, terribly wrong.

The waves began to chop higher, and at some point the launch began to struggle. A quick look aft would have shown the sub beginning to wallow, a sluggish side-to-side roll that seemed to grow worse with every passing moment. Not long after that, the crewmen would have watched in dread as the towlines began to tauten and creak, stretching before their eyes. The boat was somehow taking on water – maybe through an improperly fastened hatch, a valve left open or even perhaps a popped seam – it could have been any one of a hundred possibilities, but she was going down and there was nothing anyone was going to be able to do about it. Worse still, there was every possibility the sub would take the launch down with her if someone didn't do something – *now*.

The launch captain's options were almost nonexistent. Pressing on to Fort Morgan was out of the question, and turning around to try and at least get her into shallower water was impossible with the growing weight that was now starting to pull the launch's fantail sickeningly close to the water. The captain certainly knew the value of the little sub, but the lives of his crew were his first priority – and with that in mind, he made his choice. A shouted order, the CHUNK of an axe, the SNAP! of a parting towline – a quick bob and roll, and she was gone into Mobile Bay, her exact location still unknown. It had been one thing to scuttle little *Pioneer*, trapped in New Orleans with an overwhelming enemy force bearing down on her. But the Mobile boat hadn't been lost in combat or even due to combat related causes. She had sunk, *under tow*, because she was unable to even proceed to her own patrol station under her own power.

The silence that night at Park and Lyons must have been devastating as McClintock, Watson and Alexander tried to figure out what had gone wrong. Baxter Watson at least tried to get some kind of recovery effort underway – according to Mark Ragan, he wrote to Confederate Navy Secretary Stephen Mallory asking for funds to try and raise the boat.[24] Mallory strongly encouraged technological innovation but apparently had doubts about submarines, so he passed the buck back to Buchanan and asked what *he* thought of the idea. Buchanan was polite and respectful but to the point: the Mobile boat was simply impractical for any kind of operations within Mobile Bay, and therefore he must respectfully recommend against any kind of salvage attempt. Secretary Mallory concurred, and that was it.

On the face of it, the operation was finished. The Mobile boat was gone, silting up somewhere out there in Mobile Bay. Hunley, who had been away on Government business through most of the construction and testing, had devoted most of his remaining funds for little return – the useless electric motor, the perambulating torpedo, the boat itself. He had no money left, or at least not enough to build a new sub that would correct the mistakes they had made with this one. McClintock and Watson had even less, and Lieutenant Alexander's military pay – if and when he received it – could afford him only the barest

---

[24] Ragan, pg. 33.

necessities. The Confederate States Navy, though interested in the possibilities of submarine warfare, was not going to provide any assistance either.

First, funds were short in any event as the Southern economy was slowly garroted. Secondly, there was an official program underway that was getting whatever funds there were. These subs – at least one, possibly more – were being built at the Tredegar Iron Works in Richmond. Why anyone was building submarines in Richmond is a bit of a mystery, but never mind. It seems reasonable enough to assume that since Richmond had manpower, rail connections, and materials in relatively abundant supply and under reasonably safe conditions, building them there made sense to someone. In the event however, they didn't even have the limited success that *Pioneer* and the Mobile boat had enjoyed.

None of them ever seem to have gotten into the water, though Rich Wills suggests that the submarine pulled from the fetid waters of St. John's Bayou in 1878 may have been an experimental boat from the Tredegar works.[25] In any event, there was no more money or official support to be had to resurrect the project – Admiral Buchanan's letter, though written without rancor or bitterness, had seen to that. So, it would seem, the sun went down that chilly February evening in Alabama on a deeply disappointed team of submariners. God's arm did, perhaps, reach the ocean floor, but it did not reach to the hearts and pocketbooks of the government in Richmond.

The flickering gaslights would have been extinguished, the doors bolted and locked for the night, and four men would have hunched down into their coats against the wind coming in off the bay as they trudged silently back to their rooms. Mobile, a city facing imminent siege, was quiet as workers headed home to families and friends under a clear sky. It may have been cold, but there was dinner and a warm fire to greet them. And out over the horizon, in the darkness past Fort Morgan, more ships with more men grimly standing watch would join the blockade, making it just a little bit tighter.

III: "DIVE WITH OUR MEN BENEATH THE SEA..."

The spring of 1863 was a strange time during the War Between the States, a time of tragic victories and unexpected defeats.

Robert E. Lee had delivered a stunning defeat to the Federals at the Virginia town of Chancellorsville across four days in May of 1863. US General Joseph Hooker had taken the Army of the Potomac (AoP) out on one of its periodic marches towards Richmond, and he wasn't at all shy about telling anyone who'd listen about exactly what he was going to do once he got hold of Bobby Lee. Hooker had started off well enough, apparently stealing a march on Lee and getting the AoP across the Rappahannock River just west of Fredericksburg and getting into position on either side of Lee's Army of Northern Virginia (AoNV). It was a tricky move, but beautifully executed and it put Hooker in a position to close the trap and end the rebellion once and for all. Hooker even issued an order of the day stating that the 'enemy must ingloriously fly' or face certain defeat.

Unfortunately, no one ever seems to have told General Lee. Over the next few days, Lee and the AoNV put on a display of skill and tactics that put Hooker on the defensive from the very beginning and kept him there. He never did regain the initiative; seemingly terrified that Lee was about to do to him all the things he'd planned for the regal Virginian. For some time in the battle, Hooker seemed paralyzed, even more so when a Rebel cannonball tore through his HQ tent at one point and left him unconscious for an hour. When it was all said and done, seventeen thousand Union casualties - fifteen percent of Hooker's force - were dead or wounded in the green Virginia countryside as once more the AoP limped back to its base camps near Washington.

---

[25] Coski, cited by Wills, pg. 8.

Lee's casualties were almost as bad and a larger proportion of his strength - thirteen thousand and *twenty-two* percent of his total force - but for all the horror of that number, they were all dwarfed by one in particular: Lieutenant General Thomas J. Jackson, CSA. Jackson had given a virtuoso performance at Chancellorsville and was planning to top it off with an early morning attack on May 3rd. But on the evening of May 2nd, Jackson and some of his staff rode out ahead of the Confederate lines in order to get a better feel for what they were up against. As they returned from the gathering dusk, pickets from the 18th North Carolina opened fire, believing them to be a Federal cavalry patrol. They blew Jackson out of the saddle (exactly who hit him has never been determined) with two bullets in his left arm, and he was taken immediately to a field hospital.

That Jackson survived the initial wound - and the field amputation immediately thereafter - speaks volumes for his sheer willpower and for a brief time there was hope that he might indeed make it. After all, John Bell Hood would lose a leg and have an arm crippled in the service of the Confederacy, and he survived the war as did Dick Ewell, who left a leg in the Shenandoah Valley. But not even the grim, implacable Stonewall could defeat pneumonia, which set in as he tried to recover. Eighty years before antibiotics, that was a death sentence. On May 10th, the man Lee called his 'strong right arm' passed into history, his last words a murmured request to 'let us cross over the river, and rest under the shade of the trees."

There was no time to mourn, none at all. Soon after the victory he'd paid so much for, Lee was planning a full-dress invasion of the North. His aim was twofold - first and foremost to take the pressure off Vicksburg, the Confederacy's last outpost on the Mississippi River, and secondly to take another crack at the Army of the Potomac before it could fully refit and recover and to possibly break it for good. Lee himself – not a man known for impracticality – believed it might be possible to defeat the Union once and for all.

'Defeat' in this case meaning that the Army Of The Potomac would be so badly battered and demoralized that it would be unusable in its primary mission, the defense of

Washington, D.C. At that point, it would be check and mate - Washington would be exposed to a raid at the very least, and at best...even Lee, a man not known for anything except the most realistic and concrete of plans, had to have entertained the vision of gray columns riding down Pennsylvania Avenue towards the Capitol and its unfinished dome. The plan was sound, it's reasoning solid, its aim sure, but Lee would have to do it without Stonewall. Marse Robert would not realize how great a loss this was until he stood before a little Pennsylvania town called Gettysburg.

And at Vicksburg, the screws were slowly tightening just a little bit more every day.

A parade of Union commanders had made one attack after another, mostly by trying to run the Confederate batteries there that dominated the river. All that did though was leave Union ships adrift and sinking in Old Man River and Union sailors dead or dying in his cold embrace.

Finally, in October of 1862, a new commander tried his luck at subduing the little Mississippi town. He had had an unremarkable career in the old Army twenty years before, only to return home and fail at just about everything he'd tried since. Well, perhaps 'fail' was a bit too strong a word . . . he just hadn't truly succeeded at anything. When the war started, he felt himself competent to command a regiment, no more than that and said so. But now, looking down at Vicksburg, Ulysses S. Grant commanded forty thousand men, and he was quietly determined not to fail at this.

First he tried a basic, straightforward march on the Gibraltar of the Mississippi, but cavalry raids led by Nathan Bedford Forrest and Earl Van Dorn cut his supply lines to ribbons. Grant had no choice but to fall back on his main base at Memphis. He was a fast learner though and the next time, Union cavalry would protect his flanks and he would live off the land itself, cutting himself free from a fixed base. He then made four attempts to dig his way around Vicksburg, trying to make canals that could be used by Union gunboats and shipping. These failed too, but not through lack of trying. And as hard as it was on the poor

CSS H.L. Hunley

bloody infantry that suddenly found itself employed as ditch-diggers, they were developing a genuine affection for the quiet, modest general who refused to throw their lives away on one frontal attack after another.

In the early spring of 1863 Grant tried once more and this time he went for broke. First the Navy was finally able to run the batteries on April 16th, catching much of the Confederate leadership at a gala ball that night. Then Grant moved out, methodically knocking Vicksburg's supply lines out from under it. By May 19th he was within sight of the fortress and then finally ordered an all-out attack. It failed to take the city - as did a second attack on the 22nd - but the Federal lines were now close enough that all traffic in and out of the city was interdicted, and the awful word 'siege' started to be whispered in Vicksburg's defenses.

Now reinforced, Grant had seventy thousand men surrounding the city and he knew he could wait. The Confederacy had no more men, no more weapons, no more supplies to send and it was simply a matter of time. Vicksburg's commander, General John Pemberton (who had once, on behalf of Captain Robert E. Lee, officially commended Lieutenant Ulysses S. Grant on a job well done) his soldiers, and the civilians under his protection would find themselves eating rats and suffering from scurvy as it slowly began to sink in that there would be no rescue.

On June 28th, a fair number of Pemberton's own men signed a letter suggesting that surrender - instead of slow starvation and mounting desertions - should be considered. Pemberton's own staff was honest and even more blunt: in their weakened, starving state a breakout attempt was not even an option. So, on the late afternoon of July 3rd, Pemberton sent out a detail under a flag of truce. Grant entered the city on July 4th, a holiday that would not be celebrated again in Vicksburg for nearly eight decades.

It was also the day after Robert Lee's men, unable at the end to take just one more ridge, were marching back home from Gettysburg.

The Long Patrol

The Confederacy was now - short of an increasingly unlikely outside intervention or Federal military collapse - doomed to a slow strangulation. The Mississippi River, in Lincoln's magnificent phrase, "flowed unvexed to the sea", and the United States Navy was its master. The blockade had finally taken hold and the odds were now catching up with the daring runners as more and more were either captured, sent to the bottom, or desperately run aground for their bones to bleach on the sand. And now, a line of Blue stretched from Norfolk in the east to Cairo, Illinois in the west, a line that would be bent or even buckled in the future but never again broken. It was then that old Winfield Scott's anaconda began to stir, then coil, and finally squeeze the life out of the Confederate States of America.

We've seen, in more recent wars the desperate attempts of our adversaries to find something, *anything* that will turn the tide in their favor. Germany imagined itself deploying 'wonder weapons' that they hoped would send the Allies reeling westward in panic, begging for an armistice. Imperial Japan hypnotized itself into believing that it was winning World War II in the Pacific. Every ship Japanese pilots was considered sunk, every airplane they shot down transmogrified into hundreds more - until US warships started shelling Nippon proper. Then it sent thousands of its surviving pilots to crash themselves into American ships utterly convinced that we could not stand up under such an onslaught. They were told they were emulating a heaven-sent typhoon a millennium before that had destroyed a Mongol invasion, a 'divine wind' - the *kamikaze*.

The history books tell us what happened, who won, and why. The "wonder weapons" were far too little and too late (and simply not good enough) to stop the *Gotterdammerung* that was bearing down on the Third Reich. Perhaps one percent - and probably much less - of the kamikaze force got through to their targets. They did horrible damage and inflicted immense casualties, but their only real accomplishment was to help convince the US leadership that the only way to deal with such a people was to use the atomic bomb.

The Confederacy was neither Germany nor Japan. They sought neither conquest nor extermination, only to prove their unquestioned and rightful independence and be left alone. But they must be numbered among the True Believers, those who fought on to the end believing to the last tick of the clock that they could find something, some Excalibur that would drive the invader back and forever freeze time to that haunted April of 1861. The sheer, brutal military logic of the situation would always cast a shadow over the Confederacy, and that shadow would eventually become a shroud.

But that wouldn't stop a great many people from trying.

In Mobile, the blockade was a weary, depressing reality; a wall of wood, iron, sails, and rigging that was becoming more impassable by the day. Because of its physical geography, Mobile Bay was fairly easy to block, and by the summer of 1863 it's unlikely that anything useful was getting through in any practical quantities, though runners would continue to make for Mobile until the bay was finally sealed in August of 1864. That didn't change Mobile's strategic importance, far from it. It was a given that one day the United States Navy would shoot its way past Fort Morgan, and that would be that. All General Dabney Maury could do was make the effort as expensive as he could for the Federals and by now he'd done quite well.

General Dabney H Maury, Confederate States Army. Source: U.S. Army.

Mobile Bay had been turned into one huge water-filled trap, intended to make any aggressor pay a high price for his impetuosity. There were eight batteries - not counting Fort Morgan - that commanded the entrances to the bay as well as the approaches to the city itself. Several lines of 'electric torpedoes' - command detonated mines that could be fired from shore - were laid throughout the bay, and several large moored torpedo fields were sown at strategic points as well, and that is where we come back to McClintock, Hunley, et al, and the Park and Lyons machine shop.

Those torpedoes were designed and manufactured by a gentleman named Edgar Singer, late of the great state of Texas. He and a group of former artillerymen had devised a simple, reliable and - most importantly - cheap torpedo for use in defending Confederate harbors. Remember now that these are what today we call mines. The self-propelled, fifty-mile-an-hour monsters that we know today wouldn't be invented for another twenty years, when an Englishman named Robert Whitehead built the first practical ones - huge, wonderfully Victorian-looking beasts that would set the pattern for all that have followed since. (His company survived well into the twentieth century, and became part of the Vickers Company and then the Marconi Corporation, which, as part of the international Thales group, today still builds torpedoes for the Royal Navy.)

Now, Mister Singer felt that his weapon had far more potential than to simply hang about Mobile and other Confederate ports waiting for some hapless federal warship to run headlong into it. But for the time being, the first priority was to produce the torpedoes the CSA had already contracted for, and he needed help. His original team was smart and capable, but they were going to be rapidly overwhelmed by the sheer amount of work they faced. Singer needed more men to assist - and almost as importantly, invest in the project. As it turned out, there were a handful of men at the Park and Lyons shop who were temporarily at loose ends and had some previous experience with submarine torpedoes. And so McClintock, Hunley and the rest of their band joined Singer.

It wouldn't have taken long at all for Hunley and the others to start discussing their previous efforts, both submersible and explosive, with Singer and his team. Singer, no doubt, would have been very interested. After all, regardless of the success of their boats, they had produced a series of practical, reliable towed torpedoes that seem to have worked every time. From there, discussing a new underwater torpedo boat would have followed as night follows day.

CSS H.L. Hunley

By the time the discussions were finished, we could add one more title to Horace Hunley's roster of accomplishments: salesman. Hunley managed to convince Singer and three of his original partners to ante up $10,000 CSD to finance and own shares in a new submarine torpedo boat, while Hunley came up with another $5,000 CSD - and in any currency at that time, fifteen thousand dollars was a great deal of money. (By way of comparison, USS *Housatonic*, the ship *Hunley* would sink, cost just a bit over a quarter million dollars in 1861.) Hunley's team had to be grateful, perhaps ecstatic that they were going to get one more chance to prove their idea.

They almost certainly had to know it would be their last chance, for if the new boat failed there would be no more money from any source. The Confederate economy was poised on the brink of total collapse. Hunley was still the wealthiest of the group, but by that point it was a purely relative thing - he had paid for the entire *Pioneer*, lost everything he had in New Orleans, then covered most of the cost of the failed *American Diver/Pioneer II*. He was still officially employed by the Confederate government to try and find ways through the blockade, but Richmond already had a reputation as a poor paymaster. No matter what, Richmond would not provide any more money - a failed boat would end any possibility of new investors, and Hunley wouldn't have had enough left to finance a new boat on his own. The new boat had to work, and work the first time.

Exactly when construction commenced on the third boat is unknown, but it was quickly. By all accounts, James McClintock and William Alexander were the guiding lights of the new project, ably assisted by Baxter Watson and George Dixon. Hunley was often away on CSA business, but stayed in close contact with the Park and Lyons team.

Modern submarines are born in massive building sheds in places called Groton or Newport News, midwifed by cranes a hundred feet high and baptized by pinwheels of sparks from plasma welders as men assemble rings of the strongest alloys known to science. Once the hulls are assembled, the specialists arrive. The 'nukes'- the acolytes of the Reactor, the submarine's heart, harnessing the power of the stars themselves to bring the boat alive. The bookish cybernetics engineers, installing computers that can process billions of lines of code per second and turn the boat into a creature that can almost think on its own, follow them. The sonar engineers install their wizardry, hypersensitive listening devices that can detect the gentle cry of a whale's calf or the whirr of an enemy's propellers fifty miles away.

When all is done, it takes almost five years from keel laying to christening. The boat will glide out from Poseidon's forge surrounded by bunting, flags, speeches and bands, then step forward to take her place in the line.

McClintock's men had none of this. They had an early-Industrial Age machine shop - and not a particularly well-equipped one, either, one that was under blockade in a city whose defenders could look out to see enemy ships cruising in lazy circles. They didn't have five years - the Confederacy itself would barely last four, and no one knew how long it would be before the thud of heavy guns in the distance signaled the last moments of Mobile's freedom.

The technology available to them in terms of what they could make the boat do was not terribly advanced from that which had armed and equipped the first ships of the Republic eighty years before. And no one had the time for speeches and flags, not at all. The speeches and flags had belonged to that glorious spring of 1861, not the grim reality of siege, defeat, and privation. James McClintock needed to build a killing machine with what he had and he needed to build it now.

They started with a boiler, just as they had with *American Diver/Pioneer II*. What it came from is unclear - its approximate original dimensions of twenty-five feet long by forty-eight inches in diameter seem to make it just a bit too small for a locomotive, but not at all unusual for a steamship, or perhaps a building. It was cut in half horizontally, and two iron sheets roughly a foot wide were riveted into the gap. The boiler's perfect circle was now an oval roughly four feet across and five feet from overhead to bilge.

CSS H.L. Hunley

It sounds far roomier than it is. Go to the Lasch Restoration Facility in Charleston some time where the iron boat now rests, and try fitting inside the simulator that has been put together to give you an idea of what it was like inside that claustrophobic's nightmare.

The cutting and hammering and riveting would have been almost nonstop inside the brick walls at Water and State streets, from dawn until dusk and probably past that many nights as the new boat slowly morphed into the shape that has become so familiar over the years, like some monstrous sea creature emerging from an egg-case. And even as the noise died away and the exhausted workers trudged home into a sultry Alabama night, McClintock and his men would have gathered around drawings in flickering gaslight to see what they could come up with to make their boat just a little bit better.

They designed remarkably streamlined casings that were riveted to either end of the boat, and then closed off with partial bulkheads inside to create ballast tanks. The bulkheads, however, didn't extend all the way up - a design flaw that will have consequences later. Topside, two hatches were emplaced, one at either end of the boat with well designed, tightly fitting covers. The hatches used a well-designed locking system that used a long rod with a handle at the bottom, which was apparently pulled down and turned 90 degrees in order to secure it.

As the hatches themselves weighed more than one hundred and seventy-five pounds each, this was probably more to seal them against water pressure than to actually bring the hatch down.[26] Another strip of iron, this one thirty feet long by five inches wide, reinforced the top of the boat.

Now it was starting to look like the boat we know, sitting on its birthing blocks inside the machine shop and lit in flickering shadows by gaslight and ironworkers' forges.

And each day, as it inched towards completion, more news made its way in from the outside - news of Gen'l Pemberton's

---

[26] The exact configuration and operation of the hatch mechanism is still uncertain at this time - MJK

gallant stand at Vicksburg (from whence he shall surely be relieved!). News of Yankee forces fleeing headlong from the Army of Northern Virginia as it lanced into Pennsylvania, seemingly just a stone's throw from abolitionist Washington.

Massive iron ballast castings were bolted to her keel, in theory able to be jettisoned by wrench from inside the boat. Two diving planes - one on each flank, five feet long - were linked to a set of controls forward. A thick cast-iron crankshaft ran almost the length of the boat, to be turned by eight crewmen seated along the port side. At the aft end, a gear and flywheel assembly helped make for a smooth, steady ride from the heavy, three-bladed propeller.

The work pressed steadily onwards, rarely slacking as they inched towards completion. And every day there was more news, news of how Pemberton and his men were preparing for their final stand, ready to make the invader pay for every inch of Mississippi soil

News of...well, there was no news from Lee, but surely that meant all was well and any day now they would hear word of final victory. After all, Jeb Stuart and his cavaliers protected the army, and Jim Longstreet rode at his side - victory was assured, was it not?

A snorkel – two hinged iron pipes running through an air-box just aft of the forward hatch - was installed to provide some kind of air supply for the brave crew. It had a bellows rigged to it, and there are indications that one crewmember may have been specifically assigned to handle it. An effective and accurate mercury depth gauge was mounted on the hull by the forward bulkhead and above the tiller that was connected to the rudder by a hinged iron rod that ran down the deck, then up the starboard side and overhead exit through her fantail.

The captain would have his own small seat, really not more than a small plank, but it enabled him to sit more-or-less comfortably while he controlled the boat, looking out through the forward deadlights. (Apparently the boat could be conned while standing in the hatch, but this would turn out to have its own set of dangers.)

The skipper would sit with his right hand on the tiller and his left on the diving plane controls, looking remarkably as if he was flying a modern day helicopter. Manual ballast pumps were installed, one forward and one aft. Just ahead of the tiller was a compass, gimbal-mounted inside a box. To his left would also be the forward ballast tank controls, a heavy brass valve assembly

Pemberton's men ran up the white flag on July 3rd, surely due to Pemberton's northern birth - had they not been told time and time again that Vicksburg could never be taken? Lee's men were also marching home after having come so very close, their hands on the Federal cannon themselves before falling back, at the end just one ridge too many.

Stuart had - by bad luck, poor planning, and flat out arrogance - managed to miss all but the last day of Gettysburg, where he would be defeated by a young general named Custer some miles east of the actual battle. And Jim Longstreet had stood next to Lee every minute of the fight, realizing too late that Lee's final plan to crack the Federal line on Cemetery Ridge would lead to glorious, noble failure...and find himself blamed for that failure by generations of Southrons who only knew that the Confederacy had been but an arms' length from victory.

And almost every night now the sound of cannon fire rumbled in with the breakers from out past Fort Morgan, and the next morning flotsam and jetsam and splintered gray boards that had once been sleek blockade runners washed ashore in Mobile Bay.

Horace Hunley wasn't there to see his boat slide into the Mobile River sometime around the middle of July, 1863. He was again away on Confederate business, but McClintock, Alexander, Watson, Dixon, Singer and the rest of the combined teams were there to watch her slide down improvised ways into the Mobile River at Theatre Street. They would have had to figure out some way to drag the boat roughly half a mile to the dockside there, and it must have been a sight watching the boat slowly wind its way down Water Street.

There wouldn't have been a crowd watching, perhaps just a

few people who were wondering where the odd procession was going and then politely applauding when they found out its purpose. There would have been no speeches, no bands, and not even a single flag to fly over her as she slipped into the water, bobbed a few times and then settled down.

Trials started soon thereafter, probably with McClintock in command - he had been her primary designer, he had helped build her, and of everyone there he probably had the most actual time inside such a contraption. Likely Alexander and Dixon were of the party as well, along with six more men to round out the trials crew.

The first few voyages would have been simple jaunts up and down the river to get a feel for how she handled, and she handled magnificently. The handcrank-and-propeller rig seems to have been quite effective at getting the boat moving forward with ease. CSN observers estimated her to be capable of 4-5 knots, but there is no indication of whether or not this was a cruise speed or what she was capable of with her crew going all out. She steered quite easily, while diving and surfacing were simple, direct, and most importantly, reliable procedures. In short, her behavior almost immediately erased the bad taste that *Pioneer II/American Diver* had left in everyone's mouth.

The men who built her and invested in her had no doubts about her abilities, but there was one man in particular they wanted to prove a point to. So it was on the last day of July, 1863, Mobile's senior military staff stood on the shores of the Mobile - including Admiral Franklin Buchanan, CSN. Out in the river itself, carefully moored out of the main channel, a battered old scow had been moored and was now gently bobbing in the current like a sacrificial goat unaware of its fate.

Back upriver at Theatre Street, James McClintock stood in the boat's forward hatch, peering downstream at his target. This would be the boat's ultimate test and as the saying goes, failure was not an option. This would be no sprint-and-drift approach here, no careful navigation by periscope - this was going to be a direct, brute-force approach to a brute-force problem, and McClintock had neither the technology, time, nor inclination for

CSS H.L. Hunley

finesse.

The boat bobbed slightly as the crew squeezed in and took their places, illuminated only by the dim shafts of white/yellow sunlight that arrowed through the tiny deadlights. Once everyone was in place, McClintock gave the word and his shore crew pushed them away from the dock. They did so gingerly, for drifting serenely behind the boat was a torpedo carrying one hundred pounds of black powder and now studded with contact detonators. The torpedo, in turn, was attached to the boat by a line roughly about one hundred and fifty feet long.

As the boat's crew started to turn the crank, the propeller turned brown river water into white froth. It took a few seconds for sheer muscle power to overcome inertia, but as it did the boat began to ease forward, helped by the current. The wake made the torpedo bob in the water a bit but it remained blessedly unmoving as the line began to pay out, then stretch, tauten, and finally pull it forward and away from dockside.

McClintock secured the hatches and ducked down into the boat. It wouldn't have taken long for it to get stuffy, and when added to the heat of a black iron boat in an Alabama summer, the conditions inside probably deteriorated quickly. At best, there was only about three hundred cubic feet of air inside the boat, and nine men exerting themselves to that extent would have depleted it quickly. There was no time for concern there, however - they had work to do. McClintock ordered the ballast valves opened, and the boat began to settle in the water. The sound of water pouring into the ballast tanks would have echoed noisily in the boat's interior, but it may very well have been drowned out by the exertions of the crew as they pushed the boat along. The little bit of light coming in would have faded to a dull glow as she eased deeper into the river with every turn of the crank, with McClintock now half-crouched, half-standing in the forward hatch, peering through a deadlight at the target ship.

When the water started lapping up at the base of the deadlight, McClintock gave the order to shut the ballast valves. The boat was now riding awash, water covering the top of the hull and with only the two hatches and the air-box above water, with the torpedo cruising dutifully along behind it. They lit the candle now, but in all honesty it didn't help much. The CO^2 levels would have already risen dramatically, and at best there would have been only a pale glow forward and near-Stygian darkness everywhere else. McClintock would have still needed to squint to see the gauges and the compass and even then it would have been difficult to see them.

They glided along for a few more moments, and then McClintock ordered 'all stop'. Heavy as the boat now was, it still drifted a bit as McClintock tweaked the controls as best he could to line up for the attack run. From a vantage point perhaps about three inches above the water, the target ship would not have been all that visible - most likely the masts would have been poking up above the water but no more than that.

The boat was now positioned about as well as it would ever be, and it was time to go in for the kill. Ordering the crew to turn the crankshaft with everything they had, McClintock opened the ballast valves and took her down. The boat's buoyancy was always delicate to begin with, always on a knife-edge of just awash or all the way under, and it didn't take much for the hatches and air-box to disappear under the surface leaving only the torpedo chugging merrily along by itself towards the weather-beaten gray/black sides of the target ship.

McClintock was now performing a juggling act unlike anything in naval history. Half standing, half crouching in the forward hatch, his right hand was on the tiller and his left hand on the diving plane controls trying to keep the heavy boat level and straight, while his eyes were locked on the barely visible compass dial, keeping it from deviating so much as a fraction of an inch.

With no radar, no attack scope, no sonar, that needle was the only way McClintock had to keep the boat on course. Even though the compass had - in theory - been adjusted to compensate for being inside a metal tube, there had to have been a nagging doubt in his mind every time they closed the hatches and went down.

The rumble of the crankshaft and the crew's labored

breathing filled the boat, but McClintock was probably holding his breath as he kept his eyes flicking between the depth gauge and the compass needle, knowing only that they were headed closer, closer still -- And at the very moment that they must have thought that they were overshooting the target, the boat didn't quite stop short but rather gave a hitch, just long enough for McClintock's heart to skip a beat to wonder if he'd somehow bounced the boat off the muck on the bottom of the river. But almost before he could form the thought, there was a dull thud behind them and almost simultaneously, a shockwave grabbed the boat and shook it like a rat in a terrier's jaws.

Topside, the view must have been unforgettable. The silver V of the torpedo's wake arrowed directly for the target boat, followed by a sharp *crack* and the white smoke and sparks of a black powder explosion. Water, mud, stunned fish, and the shredded remains of the torpedo body fountained up into the air, all of it cascading back down into the semi-circular shockwave that was now spreading back from the explosion to wash over the riverbank. The smoke and debris would have obscured things for a moment, but when all was calm again, the observers saw the old scow lean drunkenly to one side and settling as the Mobile River claimed her.

There had to have been cheers and hurrahs, praise that only grew louder as the boat slowly emerged near the sinking wreck and James McClintock popped the forward hatch, almost jumping clear of the boat as he whooped in sheer joy. There was no possible doubt now – the new boat worked, and worked magnificently. Her display before Mobile's guardians had been powerful and convincing, and none seems to have been quite as convinced as Franklin Buchanan. Old Buck may never have completely overcome his dislike for submarine warfare, but he couldn't deny what his own eyes had witnessed. From that moment on, Admiral Buchanan was a believer.

The men of Park and Lyons would not have been human if they hadn't celebrated that night with the happiness of the saved. Not only had they proven their idea at long last, they had given their Confederacy a weapon unlike any other – combining the tough black hulls of the ironclads with a degree of stealth unsurpassed in naval warfare. Of course, any spoilsports present might have pointed out some potential difficulties that lurked in the boat's future.

First, all of her tests so far – including this most spectacular one – had been conducted in calm, protected, and relatively shallow water. There had been no tests in open water yet, though given the ominous presence of the United States Navy just offshore, this can be excused. Secondly, the boat's designers still had no solid idea of her practical endurance, either surfaced or submerged. A few short out-and-backs were more than enough to get a feel for the boat, but what would happen when the targets were on the blockade line, four to five miles offshore, in less than ideal conditions? Would the crew even be able to get the boat far enough out to make an attack run? Third and finally, how well will the boat hold up under fire? Assuming the crew could even get it that far out to sea and execute a successful attack, with a sinking target only a few yards away and other blockaders rushing to its aid, would a physically and emotionally drained skipper and crew be able to get clear and back to their base?

No one – not McClintock, not Alexander, not Singer, not anyone else on the team – could answer those questions that muggy July night, and frankly I do not think they were inclined to try. For the time being, there was no war and no blockade, just success and the promise of profit and glory. There was only the black iron monster tied up to the Theatre Street docks.

And it *worked*.

Hopefully the celebration wasn't too riotous, as it was back to business the next morning, August 1st, 1863. Everyone who had witnessed the test the day before was present for a meeting at General Maury's HQ, as well as McClintock, Singer, and the rest of the team.

As is the way of such meetings, it was a good-news, bad-news affair. The good news was that, as had been anticipated, the boat had met every single one of its expectations. It was beyond any question a viable, practical weapon that could only serve to the further benefit of the Confederacy. The bad news, however, was that it would not serve in Mobile. A single bugaboo from the late and unlamented *American Diver/Pioneer II* had come back to haunt the new boat: it needed fairly deep water to operate safely and efficiently, and Mobile Bay was simply too shallow. It could perhaps have been operated from Fort Morgan, which sat directly on the ocean. But tied up there it would have been terribly vulnerable to marauding blockaders, and almost twenty miles away from any decent repair and refit facilities.

Fort Sumter made Charleston the very symbol of Secession. Union Honor demanded that the city be punished accordingly. Source: U.S. Navy History Center.

How the decision was made to send the boat to Charleston is unclear. There was certainly a military necessity for some kind of help: since early summer, the five primary forts guarding the city – Moultrie, Sumter, Beauregard, Putnam, and Shaw – had been under round-the-clock attack by the US Navy. So far, a combination of poor Federal tactics, sheer Confederate stubbornness, and few lucky breaks had kept the Federals at bay. But if the US fleet could force either Moultrie or Sumter – and they'd gotten close – Charleston Harbor would be closed,

The Long Patrol

leaving only Wilmington, North Carolina, as the only free Confederate port on the East Coast. Mobile Bay, because of its geography and the array of weapons that was strewn through it, was probably as secure as it was ever going to get. The unique abilities of McClintock's boat were needed at Charleston, and that was where they were headed.

Accordingly, Baxter Watson and one of Singer's men, B.A. 'Gus' Watson, were chosen as the advance men for history's first forward deployment of a combat submarine. They got priority on a train out of Mobile, and arrived in Charleston sometime in the next day or so. In the meantime, everyone else was involved in trying to figure out exactly how to get the boat to Charleston in the first place. Taking her out as deck cargo on a runner was briefly considered, but not any longer than that – there was simply too great a risk of losing it. That left the railroad as the only practical option. Somehow they would have to get it from Theatre Street to the railway station, but it was going to be a challenge: the boat was thirty feet long and weighed at least six thousand pounds – not terribly heavy for the lifting gear of the day, but damned awkward.

At length, the boat was hauled back up the jerry-built ways that had been laid down for it, and checked for any damage that may have been incurred during the tests. It was then put onto a wagon and taken to the Mobile freight yards, where it was transferred by crane to the train. As the standard flatcar of the time only had a 20' bed, it was decided to couple two flatcars together, then lash the boat to them. This would make for a long, slow trip – almost three days to Charleston if everything went well, but it was a pretty reasonable solution.

The combined McClintock and Singer teams had done it: they had built a practical submarine (albeit one of limited endurance and speed, but never mind), dove beneath the sea, convincingly sank a target, and made it back. In the seven months to come, however, the men rattling eastward to South Carolina would have ample opportunity to reflect that all that had gone before was, in fact, the easy part.

Now came the trial by fire.

IV: "...TRAVERSE THE DEPTHS PROTECTIVELY..."

The ride from Mobile was long, slow, and hot. More than a few times they were shunted to a siding as troop trains chugged past, carrying scared young boys and canny old veterans in a pastiche of tattered uniforms, drab civilian clothing, rifles and swords.

The Confederacy was slowly, painfully beginning to contract that summer. It would be almost two more years yet before the Lost Cause gave up the ghost, but with the twin defeats at Vicksburg and Gettysburg the end was inevitable now. True Believers, though, never give in, never stop the fight until they are exhausted in a corner at the very end, and the McClintock/Singer teams were all True Believers. There was a chance, a slim chance glimmering in the mounting dusk, that they could make the victory so painful and so costly that perhaps...perhaps...

The train finally puffed into Charleston three days later, its passengers limp with sweat and blackened with soot and cinders, but all in one piece. That, however, would soon become a debatable proposition. Charleston – the first city of the Palmetto State, the birthplace of Secession itself, a city so filled with churches and so sure of its role and place in history that it is known to this day as the Holy City – was under fire.

Only two American cities have ever had the grim distinction of undergoing prolonged bombardment and siege – Vicksburg, of course, and Charleston. Charleston of course was never completely cut off as was Vicksburg, and remained a more-or-less open port until Confederate forces finally evacuated the city one step ahead of William Sherman's forces on February 18th, 1865. But from the beginning of the war, Charleston held a grim fascination for the Federals, perhaps second only to Richmond itself. Rare is the American who hasn't opened a history book to see the Charleston *Mercury's* famous front page, "The **UNION** is **DISSOLVED**!!!" At the time, there were many published reports – and more than a few illustrations – of the good people of Charleston strolling, picnicking and cheering like fairgoers on the Battery during the thirty-six hour bombardment of Fort Sumter. And last of all, but hardly least, the sleek gray blockade-runners that still slipped past the Federals were one more gesture of defiance that couldn't be tolerated.

The Scourge Of Charleston: Admiral John A. Dahlgren, the 'Father Of Naval Ordnance' and commander of the South Atlantic Blockading Squadron aboard USS Pawnee. The official caption for this photo states that it was taken between 1863 and 1865, but the latter date is more likely, and after the city was occupied - the object on the horizon above Dahlgren's shoulder is Fort Sumter, and were this during hostilities it is reasonable that the setting would be nowhere near as composed. (Photo courtesy of the United States Naval Historical Center, Washington, DC.)

The man in charge of making Charleston pay was Rear Admiral John Dahlgren, USN, and the commander of the South Atlantic Blockading Squadron. Dahlgren had made his name as a weapons specialist, perfecting the powerful, accurate, and deadly cannon that bears his name. His son Ulric took another path, joining the US Army and becoming a Cavalry officer. He would die a glorious death while raiding near Richmond, and suffer a sad afterlife as his corpse was strapped into a coffin and displayed as a war trophy. Of such things are Civil Wars made.

In any event, Dahlgren had a mixed bag of ships at first, but as the effort began to pick up speed, the toughest ships in the growing US fleet were assigned to the effort to crack Charleston, which at least in terms of fortifications was the best defended port of the Confederacy. The new ironclads were being sent to Dahlgren's command as quickly as they could be turned out, which was a mixed blessing for the Admiral. On the one hand, they were the only ships that had a chance of standing up to the heavy guns at Sumter and Moultrie and in turn punching through their heavy masonry fortifications. On the other hand, building and designing them was still more art than science. The ironclads were slow, under-armed (usually only one or two heavy caliber guns and a handful of smaller ones), unmaneuverable, and often unstable. One, USS *Weehawken*, rolled over off Morris Island later that summer and took thirty-one of her crew with her after a stiff breeze pushed a few small waves over her bow. Improperly trimmed, she filled and capsized almost before anyone knew what was happening.[27]

Dahlgren's opposite number was, as it turned out, General Pierre G.T. Beauregard. After the opening round in Charleston in April of 1861, Beauregard had been eager for action and had ended up at First Bull Run, then out in the Western Theatre. He performed competently at Shiloh, but then landed in trouble for evacuating seventy thousand troops out from under the nose of Union General Henry Halleck at Corinth, Mississippi. In fairness to Beauregard, his troops were in no shape to make a stand, many down with dysentery and typhoid – trying to stand up to Halleck's comparatively healthy, well-fed and well-supplied troops would have been noble, but futile in the end. Beauregard didn't make the bitter pill of the retreat any easier to take by his

---

[27] Cussler, *Sea Hunters II*, pg 145

grandiloquent statements on how he was going to take the place back, and then – to top it off – he went AWOL in the process of recovering from his own illnesses. CSA President Jefferson Davis had had quite enough by that point and relieved him of his command. Beauregard was something of a placeholder for a bit before being sent to Charleston in November of 1862. He relieved John Pemberton, who went on to meet Ulysses Grant at Vicksburg a few months later.

Beauregard's flamboyant persona was perfect for his new assignment – he was looked up to by the citizenry as a true Southern cavalier, there to defend them against the blood-crazed Yankee invader. He was also fondly remembered there for his actions in starting the war in the first place. For his part, Beauregard reveled in it. Whatever shadows his ego may have cast across everything else, his personal courage and integrity was unquestioned and he threw himself into the job, upgrading the fixed defenses around the harbor and personally standing up under fire.

General P.G.T. Beauregard, Confederate States Army. Source: U.S. Army

The main defenses of Charleston Harbor that Beauregard had to work with were built around three primary fortifications. Oldest – and smallest – was Castle Pinckney, built on a small island towards the outer edge of the harbor itself. A small circular fortification of brick and mortar, it was built just before the War of 1812, and like so many coastal fortifications of that time eventually abandoned once the threat had passed. However in this case there was a good reason for it – the strong and beautifully built fortifications had begin to settle and sink almost as soon as they were built. As early as the 1820s there are reports of high tides flooding through the lower casemates and into the castle's interior.

In 1832, it was manned again, but not by the US Army. Rather, South Carolina militia troops took possession of it during the Nullification Crisis – a grim harbinger of what was to come in 1861. Congress had passed a series of tariffs that had badly hurt the Palmetto State's economy, and as the situation worsened, the rhetoric became more and more heated, topped finally by John C. Calhoun's resignation as Vice President.

Returning to South Carolina, Calhoun took the lead in proclaiming that states had the right to ignore – or nullify – any Federal legislation that it disagreed with. Sure enough, a state convention proclaimed the tariffs null and void within the borders of the Palmetto State, and talk of secession and even war began to be heard. Andrew Jackson, President at the time, was not the sort of man who took threats lightly and sent the Army and Navy to Charleston Harbor to remind the residents who was in charge. In 1833, Congress passed a law authorizing Jackson to use force if need be to enforce the tariff. Things remained tense until Henry Clay negotiated a face saving compromise for both sides allowing for the tariff to be slowly reduced. Just as well – Jackson was certainly not going to back down, even if it meant firing on American citizens, and Calhoun couldn't back down even though he had been unable to find another state to follow his lead.

With that, Castle Pinckney slumbered again until the first rumblings of the Civil War. A lone US Army Sergeant, his wife, and his daughter comprised the entire garrison when South Carolina militia arrived on December 27, 1860, to take possession of said property. There is a marvelous woodcut of the event, showing the Militia storming ashore with flags flying, and the good Sergeant standing most nonchalantly at the sally port[28]. The Sergeant, recognizing that discretion was indeed the better part of valor, surrendered Pinckney and was given transport to Fort Sumter where Major Robert Anderson had evacuated Fort Moultrie's garrison just a few hours before.

---

[28] Burton, pg 176

The Long Patrol

When General Beauregard arrived Pinckney was a POW camp holding members of the luckless 79th New York Infantry, battered at First Bull Run. Beauregard had the Castle's armament upgraded, installing four columbiad cannons, seven-and-a-half-ton monsters that promptly began to settle into the sandy soil. The columbiads, by the by, remain there to this day, buried almost to their muzzles in sand, pebble, and oyster shell fill, mute testimony to the nondescript little island's past.

Next closest was Fort Moultrie, slightly north of Charleston on the eastern edge of Sullivan's Island. This was actually the third fortification to bear the name – the first two, of Revolutionary and post-Revolutionary origin, had quickly vanished from the scene. The first one simply deteriorated, and the second one was erased by a hurricane only six years after it was finished. The present fort was authorized in 1807 and completed in remarkably good time, just two years. This Moultrie was a state-of-the-art coastal fortification, with strongly built brick ramparts that were proof against even the largest smoothbore weapons of the time. And even more remarkably than the speed with which it was built, Moultrie stayed garrisoned through the years although very little was done with it except for routine upkeep and maintenance.

All that changed in a day in 1860, one day after a grim Christmas in Charleston and the last peaceful one it would see for five years. US Army Major Robert Anderson - a stern but gentle and compassionate Kentuckian married to a Southern belle, and who actually sympathized with the South but who wanted only to see the Union survive - was the garrison commander at Moultrie and he was living a nightmare. The Moultrie garrison, undermanned (approximately eighty-five men, a number surpassed most days today by the number of tourists who visit it), unsupplied, and with all its heavy guns facing the wrong way, was a sitting duck for the Confederate forces that ringed the fortifications. Moultrie then looked considerably different than it does now, and there were small hills and buildings around the fortifications that gave snipers – and artillery – a clear line of fire into the fort itself. In addition, the seaward side of the fort - today kept clear and spotless, was

CSS H.L. Hunley

at that time almost swamped by sand blown off the beach - apparently high enough that local cows had been known to wander into the fort that way.

There was, however, good news to soothe the nervous Federal troops: there was a Government commission taking care of everything, though it seems that Major Anderson took little comfort in that. Since South Carolina's secession on December 21^{st,} a group of South Carolinians (mostly former Congressmen) had been meeting with President James Buchanan to determine the status of what they considered former Federal property in the Palmetto State. Buchanan – whose spine was almost nonexistent when it came to the final showdown with the South – was actually more than willing to consider an offer to purchase said property when Attorney General Edwin Stanton (who would later become Lincoln's ferocious Secretary of War) and Secretary of State Jeremiah Black flatly told Buchanan that they would resign rather than support such a deal.

At that Buchanan suddenly found courage that, had he used it sooner, may have prevented the whole awful thing – he told the commissioners that not only was the answer no, he reserved the right to keep Federal forces in South Carolina and the right for them to defend themselves. That certainly wasn't what the Palmetto delegation was expecting, so a compromise of sorts was reached. While the matter was being hashed out, nobody would shoot at anybody else, and no attempt would be made to reinforce the Federal troops at Moultrie or move anyone in a blue uniform anywhere in what was now considered sovereign South Carolina.

The agreement - to be kind - was open to interpretation, especially by the South Carolina delegation. The Palmetto diplomats and South Carolina Governor Francis Pickens believed that President Buchanan had a gentleman's duty to keep his word no matter what – especially no matter what the Confederacy did. This was made worse by the fact that Buchanan's Secretary of War, John Floyd - a slippery character who was embezzling millions of dollars with one hand and selling muskets to the soon-to-secede South Carolinians with the other - was telling the Carolinians whatever they wanted to hear, whether it was true or not. (Floyd was expecting any time to be arrested for his embezzlement and gun running, and was trying to open a line of retreat into the Confederacy.) Buchanan however, showing his usual grasp of a thorny problem, was not at all sure that he was dealing with men who had the authority to make – or keep – *any* agreement.[29]

On the face of it, this agreement was being kept. As a practical matter, both sides seem to have been working very hard to subvert the whole damned thing to their own benefit. First on the Federal side, plans – most of them far from practical under the circumstances - were being made almost from the start to reinforce Moultrie despite official government orders not to. This seems to have stemmed from the military's universal hatred of Buchanan's actions leading up to secession and their knowledge that Abraham Lincoln was not going to simply roll over the way his predecessor had.

On the Southern side, what was happening was that every Secessionist with the resources to get to Charleston was going there, and although Beauregard was trying hard to keep them under control it wasn't working. Mobs, encouraged by some of the more fiery speakers, surrounded the fortress and generally made threatening gestures towards Anderson's men, while the South Carolina militia – Confederate regulars hadn't yet taken up posts around the fort – weren't exactly behaving in a military manner and indeed could frequently be seen having a good time with the mob, their military duties notwithstanding.

The mobs alone would have been able to overwhelm Anderson's men and for all his responsibility Robert Anderson was not going to start a civil war by opening fire on civilians, armed or not. This potential horror was compounded by a near total lack of information or instructions from Washington – except for 'sit tight, and don't open fire'. On top of that, rumors were starting to fly that there would be an attempt by the Militia to take the fort, agreements notwithstanding, and that the infant CS Navy was going to blockade Moultrie to keep the garrison

---

[29] 'Dilemmas of Compromise', *The Crisis At Fort Sumter*
http://www.tulane.edu/~sumter/Dilemmas/DDec26Comm.html

from being resupplied – or escaping.[30] Major Anderson was a man who followed orders and would have given his own life to see that they were followed to the letter – but absolutely nothing in his orders *required* him to sit at indefensible, defenseless Moultrie and allow his men to be overrun. Which brings us to the last and greatest of the Charleston fortifications – Fort Sumter.

When we think of Charleston and the Civil War, we think of Fort Sumter, that grim three-story masonry and iron wedding cake at the mouth of the harbor where four years of bloodshed, horror and glory began. But just a few days before the first shot went arcing into the spring night, Fort Sumter simply wasn't that imposing – or for that matter, all that dangerous.

Sumter had been started in 1829 on what had been a shoal in the harbor's mouth. The Army's engineers expanded the shoal to sufficient size to erect a massive fortification that would be one of the largest built on the East Coast, capable of mounting one hundred and forty guns ranging from thirty-two pounders up through awe-inspiring eight-inch columbiads and squat, wicked-looking black mortars. The problem was that no one had ever really gotten around to completing the place. The weapons were awaiting mounting in their carriages, barracks stood ready to receive gunners and soldiers, but there was almost no ammunition and even less in the way of useful supplies.

It did, however, have one advantage Moultrie didn't: the only way anyone could get a shot at it was from artillery, and Sumter was designed to survive days, if not weeks of heavy bombardment. It wasn't much hope, but it offered a better chance of staying in Federal hands than Moultrie did, and Anderson had to make the call. Anderson also had to face the possibility - not completely impossible - that the War Department might tell him to go back to Moultrie. Finally, early on the morning of December 26th, Anderson made his decision: evacuate Moultrie and get the garrison the mile-and-then-some out to Sumter.

---

[30] John Peltzer, 'The Union's Mission to Relieve Fort Sumter', *America's Civil War Magazine*, September 1997

The Long Patrol

The Major pulled it off magnificently. Early on the morning of December 26th, the garrison began their day with a series of orders at morning parade that led everyone to believe that they were getting ready to defend Moultrie to the bitter end. That suited Anderson just fine, who then called some of his officers together and told them what was really about to happen. Just at noon, the wives and children of the garrison were loaded on a boat at the Moultrie wharf and sent on their way. Some of the residents of Moultrieville noticed the movement and became suspicious, but the explanation that the dependents were being sent out of the line of fire made perfect sense, and the locals went back to their holiday routine, assuming that the dependents were being sent over to Charleston.

The old saying about what assumptions do to you and me has rarely proved more correct.

Anderson had decided to send the men over at about 5:30PM, just as it was getting dark. The sentries and the locals would be distracted by dinner and drink, and the darkness would help even more. At about five o'clock, Anderson told the rest of his officers what was going on, and the men hit the ground running. The boats - three good-sized rowboats - were hidden by the old seawall, probably at the foot of what is now Station 12 Street on Sullivan's Island. The garrison, with Captain Abner Doubleday's men going first, quietly made its way from the fort. It would have been easy to go over the wall, and with the bulk of the fort masking them from the locals they made it to the boats. There were a couple close calls with ships passing through the channel, and one patrol boat even stopped to get a look at them, but then went on its way, apparently believing that they were workers headed to or from Sumter.

When they arrived, the hundred or so civilian workers who were still trying to finish the place were rousted out of their warm beds by a grim Captain Doubleday, who told them they were done working on the fort and backed it up with fixed bayonets. The boats turned around and headed back to Moultrie to pick up the last men there. They then quietly left Moultrie but not before firing two signal guns that let the boats with the dependents - sitting quietly off Fort Johnson, about a mile due east of Sumter - know that they could safely head the fortress and rejoin their husbands. A small team remained behind to destroy the gun carriages, but they were still at it when the sun came up on December 27th, and the rebels were finally beginning to figure out they'd been had.

General Beauregard was not pleased. Governor Pickens, in turn, was almost as pleased as General Beauregard. Anderson had technically violated the agreement between the commissioners and Washington, and both sides had red faces – the Confederates for being caught napping, and the Federal Government having to admit they had no idea Anderson was going to try something like this. There was also the strong possibility that Governor Pickens wanted to take Sumter himself. The military chain of command at the time was, to be charitable, somewhat amorphous as the states that had seceded considered themselves sovereign nations until the formation of the Confederacy – and that made Pickens, as the senior elected official in the state and already the commander in chief of the Militia, the senior military commander in the state.

A few weeks later, Pickens would give the orders to open fire on the *Star Of The West*, the supply ship that would be the first desperate effort to reinforce and resupply Anderson's men – Pickens knew exactly what he was doing, and if there would have been any way to run the Palmetto Flag up the Sumter flagstaff, he would have done it and a gentleman's agreement be damned. Mind you, it was something Beauregard would have never countenanced for a number of reasons – first, he didn't have the men, and secondly Sumter could not be supplied with the resources he had available at the time. But in any event, the war came very close to starting right then and there, as the good people of South Carolina were furious and Governor Pickens was ready to let them try and take the place back – all the more so because he felt he had an ironclad agreement with Washington that nothing like this was going to happen. [31]

In high dudgeon – and with no other choice - Beauregard sent a delegation to speak to Anderson in an attempt to convince

---

[31] 'The War Begins! Fort Sumter', *Confederate Military History, Volume 5, Chapter I*

The Long Patrol

him to come back. Led by a Colonel Pettigrew, cards and proper greetings were exchanged and in a tense and formal meeting Pettigrew tried to convince Anderson that he needed to come back to Sullivan's Island. Anderson's reply was that due to a lack of communication with his government, he was unaware of any agreements between the Governor of South Carolina, General Beauregard, and anyone else. He had the responsibility of maintaining his command as a viable fighting force, and he intended to do exactly that. With that Colonel Pettigrew took his leave, no doubt turning the air blue once they'd cleared Sumter's old sally port on the South side of the fort.

Unfortunately, given the change in administrations, not a great deal got accomplished between December 26th and Lincoln's inauguration. The single effort of the Buchanan administration, the supply ship *Star Of The West*, tried to make it through to Sumter early on the morning of January 9th, and was fired upon by Battery Gregg on Morris Island, slightly to the south of the harbor. Moultrie's southernmost batteries opened up as well, and the ship was now headed into a terrifying crossfire. *Star Of The West* tried to evade as best she could, but it was too much; she had to turn around and sail out of range, just one and a half miles from Sumter.[32]

The morning after the inaugural – March 5th - Lincoln received a letter from Major Anderson that had been sent out on February 28th. Anderson was respectful but direct: Sumter's days were numbered. Even undermanned as he was, Anderson estimated that at best he had four to six weeks before he was starved out. Old General Winfield Scott was even more pessimistic; he felt that even before the provisions ran out, the Confederate Army would take the fort – possibly by March 12th.

Scott felt the best option was to follow the terms of the truce originally implemented at Fort Pickens, Florida – send reinforcements by ship but have them stand by unless the fort itself was attacked. All that did, however, was tie up precious transport hulls and leave the troops exposed to the disease and filth of cramped and indeterminate shipboard life. After a long

---

[32] On April 17th, *Star Of The West* would be captured by Texas militia – MJK.

CSS H.L. Hunley

cabinet meeting on March 29th, Lincoln asked Scott to tell him how many troops would be needed to reinforce Sumter.

Scott's reply seemed to indicate that the General-in-Chief of the United States Army had already written off the largest fortress on the east coast. Scott stated for the record that it would take up to *eight months* to put together an expedition sufficient to relieve Sumter, and that Anderson be ordered to evacuate. That in turn sent Lincoln's cabinet into a fury, and the consensus was that this was where the line had to be drawn.

Accordingly, Lincoln ordered Scott to hold the line in general and reinforce Fort Pickens in particular, the truce be damned. Pickens was indeed reinforced and stayed in Federal hands the rest of the war. The ease with which Pickens was reinforced gave heart to those who wanted to do the same thing with Sumter. One of those was a Navy officer with the imposing name of Gustavus Vasa Fox, who had come up with what was probably the most realistic plan for reinforcing Sumter, given the constraints they were dealt.

Gustavus Vasa Fox, U.S. Navy. Source: U.S. Navy Historical Center.

Fox's plan - or at least a variant thereof - was originally submitted to the Buchanan administration, but was rejected out of hand. Fox promptly sent it to Lincoln's advisors, and it was pleasing in their sight. A mixed convoy of warships, transports, and tugs would run into Charleston Harbor, surrounding a single large steamer that would carry about three hundred troops. The steamer would have to crossload the troops over to the tugs for the last few hundred yards out to Sumter, but on the whole the plan was sound, well thought out, and feasible.

The Long Patrol

After a series of acrimonious cabinet meetings, Lincoln ordered Fox to Charleston Harbor to personally survey the situation. Amazingly, Fox traveled overland and was permitted through the Confederate positions to visit Sumter. His visit was short, only about two hours – but he was able to hint to Major Anderson that there was a relief expedition planned. Not surprisingly, on his return to Washington Fox confirmed the feasibility of his plan but also reminded the cabinet that time was short – it was now March 25th, and even by the most optimistic estimates, Anderson only had about two to three weeks left. Unfortunately, at this point someone leaked Fox's report to the press.

Needless to say, all Hell erupted in and around the Holy City. Beauregard, without enough troops or resources to take Sumter but under pressure to do something, urged President Davis to negotiate the evacuation of Sumter as soon as possible. Davis, whose military experience was considerable (a West Pointer, a decorated officer in the Mexican War and Secretary of War) was smart enough to realize that was the best option. On the other hand, General Scott – who suggested the same thing – had to face a cabinet and public itching to teach the rebels a lesson. The cabinet almost immediately repudiated Scott, but it was April 4th before the final decision was made to go. Lincoln sent a letter to Major Anderson telling him to try and hold out until April 11th or 12th.

It was going to be close – Fox and the relief convoy didn't depart New York until 0800 local on April 9th. That followed a series of bureaucratic missteps and other disasters that would have seemed flatly beyond belief in fiction – there was an argument between Fox and the commander of the parallel expedition to Fort Pickens as to who would get the cruiser USS *Powhatan*, and that was followed by second thoughts on Lincoln's part as to what exactly the mission was to accomplish.

When that was resolved, the mission was now intended to simply resupply Sumter instead of relieve it. When the ships finally got underway, they were not only delayed by bad weather, the mission was now headed for Fort Pickens instead of Sumter.

CSS H.L. Hunley

In the meantime, Beauregard officially cut off the supplies to Sumter, as well as the mail. The media, ever eager to assist the war effort, was now trumpeting Fox's departure and dutifully noting his progress. The Confederate emissaries in Washington were notified officially that the expedition was on its way, and their response was predictable. Jefferson Davis, without need for intelligence work, stated his nation's position quite plainly: Fort Sumter was to be evacuated at once, else General Beauregard would be ordered to turn the fortress into rubble.

On April 11, General Beauregard sent his staff out to Sumter and delivered his ultimatum: evacuate the fort *now*. Anderson's response was quiet but firm, direct, and unsurprising – no. But in passing, Anderson quite candidly stated that they would be out of food within a few days in any event. Beauregard passed this on to Davis, who in turn gave permission for Beauregard to take Sumter however he saw fit. Beauregard, ever mindful of the Proper Order of things, sent an emissary out to Sumter to notify Anderson that the war was about to start.

They arrived at 1:30 in the morning on April 12th, and Anderson did his best to keep the game going through negotiations. Anderson made a double-talk laden offer to evacuate in three days time, but there was nothing left that could be done, and everyone there knew it. Beauregard's emissaries looked at the offer and made their reply at 3:20 AM: the batteries surrounding Sumter would open fire in one hour. There were awkward salutes and muttered goodbyes, and as they rowed across the dark water, Anderson roused his men and got them to their posts.

And at four-thirty a little bit of the world ended, and the rest of it would never be the same.

Everyone knows what happened next; but some of the details still amaze - Anderson awakened by the first shots at 4:30, and knowing his men were dug in and protected, simply rolling over and going back to sleep until 7. The men of the fort having breakfast before opening fire on Moultrie and Morris, then spending their off time making cartridge bags from Major Anderson's socks. One of the Ordnance sergeants, forbidden

The Long Patrol

from firing the heavy weapons on the upper level of the fort, defying orders and firing each one in a personal war against the Confederacy. It all rolls back through the flame and smoke, the triumphant shouts of the civilians on the Battery, the shouts and infuriated commands, into a single, long clap of thunder that was two days of Hell for the men of the fort until their surrender on April 13th. Gustavus Fox - delayed by weather, confusion, and timidity - had arrived just over the horizon from Charleston that evening, and watched it all happen in real time from just beyond cannon range.[33]

Socially, for Beauregard the city wasn't a bad assignment, as the citizens of Charleston – at least the old planter aristocracy and the wealthier merchants – still tried to carry on as if the United States Navy wasn't within sight of their luxurious homes on the Battery, and still conducted the usual rounds of balls, cotillions, and receptions. Beauregard, who loved the braid, brass and banners that went with military life, reveled in it and was a frequent and popular guest. Until almost two years later, to a very great extent that living dream was preserved. In the words of *The Charleston Illustrated*, for its military and social elite, the city remained:

"...Beautiful as a dream, tinged with romance, consecrated by tradition, glorified by history, rising from the very bosom of the waves, like a fairy city created by the enchanter's wand, Charleston affords a fit theme for poet, novelist, historian, and tourist. The family names of the Cavaliers and Huguenots still live to tell of the origin of the people; Moultrie still frowns above the bay that resounded to the first cannon of the first revolution a hundred years ago; grim visaged Sumter stands..."

The war, however, was determined to intrude on the more pleasant aspects of General Beauregard's duties. On April 7th, 1863, a Federal force of nine ironclads made a run at the city's fortifications. The result was less than what the Union commanders had hoped for.

Beauregard had done his job superbly, and the gunners

---

[33] W.A. Swanberg, First Blood: The Story Of Fort Sumter, Scribners, New York, New York, 1957

whose posts stretched from the Isle of Palms south to Morris Island were ready. The Federal task force made its attack at 2:30 in the afternoon, and they took a horrendous beating.

The ironclad frigate USS *New Ironsides*, the task force flagship, took the first of *ninety-three* hits from Forts Sumter and Moultrie and fell out of line, her steering gone. Ahead of her, the ironclads USS *Keokuk* and the doomed *Weehawken* were being hit dozens of times each, both falling out of the line soon thereafter. They were quickly followed by USS *Patapsco* and USS *Passaic*, each one holed as many as seventy times each.

It was all over by 5:30, and Charleston's defenders had decisively driven the US Navy back out into deep water. They had scored a phenomenal four hundred and ninety-three hits, averaging not quite three a minute, and quite possibly a record for fortresses against ships. *Keokuk* continued to take on water before sinking fourteen hours later, amazingly not losing a single man – the only ironclad actually lost that day.

The unlucky ironclad was the target of a daring salvage operation performed over the next few weeks, completely at night, by Confederate engineers. When it was over, *Keokuk's* twin 11-inch rifled guns were in Confederate service, and one of them can still be seen on the Battery in Charleston.

It was not the same kind of performance the Navy had turned in under David Farragut when he ran the forts below New Orleans. The Federal task force commander, Admiral Samuel DuPont, was relieved in disgrace and was replaced by Admiral Dahlgren – who resolved to do things much differently.

This is believed to be the first known combat 'action' photo. Source: Library Of Congress.

The Long Patrol

Fort Sumter following the massive August 17th, 1863 bombardment. (Photo courtesy the Library Of Congress)

McClintock and the team arrived on or about the tenth of August. On August 17th, while they were still preparing the boat for its first mission, the US Navy came calling once more. This time they came directly at Fort Sumter itself with every gun Dahlgren could muster and steadily, implacably turned Sumter into nothing more than a pile of shattered bricks. The guns that had once covered the main approaches into Charleston were utterly silenced, although Moultrie and the batteries to the south remained intact. It was still impossible to force their way directly into the harbor itself – besides the surviving batteries that ringed the harbor, there were entire fields of Mister Singer's torpedoes moored near the channel, and even Dahlgren knew better than to try and force his way through them.

In any event, Sumter wasn't abandoned – a Confederate Army infantry unit moved in to replace the artillery – but it would no longer be able to threaten the South Atlantic Blockading Squadron. Dahlgren then moved in as close as he could and started throwing rounds into Charleston proper. A U.S. Army artillery unit that had been landed near Morris Island would aid him in the assault. Through sheer brute force they had emplaced a massive eight-inch Parrott rifled cannon in the nearly impassable marshes southeast of the city and aimed it directly at Charleston.

CSS H.L. Hunley

Federal ironclad monitors attack Fort Sumter in September of 1863. By this point. Sumter's seaward guns had been silenced after the Federal attacks of August but there would still be operational batteries covering the channels in and out of the harbor until Sumter itself fell in February of 1865. (Photo from The Photographic History of The Civil War in Ten Volumes: Volume One, The Opening Battles, via Wikipedia Commons.)

The Federal artillery commander, before opening fire, did have the decency to send a message to General Beauregard demanding that he immediately surrender Sumter and evacuate Morris Island. Beauregard apparently never got the demand, but as far as the US Army was concerned, honor had been served and they opened fire on downtown Charleston with little concern for where or upon whom the shells might be landing.

Grimly nicknamed the Swamp Angel, the huge cannon opened fire at 0130 hours on the morning of August 22nd, 1863, using the steeple of St. Michael's as an aim point. Fifteen rounds were fired before dawn, some of them specially made incendiary shells. They opened fire again on the night of the 23rd and kept firing until the cannon's barrel burst on the thirty-sixth round of the day. Surprisingly little useful damage was done,

The Long Patrol

and the entire effort seems to have merely enraged the populace of Charleston and cost the US Army a great deal of money. The Swamp Angel's emplacement is still visible, though almost inaccessible, in the swamps of Morris Island while the Angel herself - burst barrel and all - today rests in a park in Trenton, New Jersey.

"...the Holy City was under fire..." Charleston, South Carolina, Confederate States of America, April 1863. The Swamp Angel's bombardment of Charleston was a sign that the whole nature of warfare was changing. (Photo courtesy the Library Of Congress)

The point, however, was made: the Federals were willing to open fire on noncombatants in order to achieve a military victory. This was not relatively quiet, peaceful Mobile, where the war – though close – was still only a rumble over the horizon. This was real war, with a real enemy shooting real shells at you. No warning other than a noise like rustling leaves that grew louder and louder and paused for just a heartbeat, then followed by a flash of light and an echoing boom that told you some other poor devil had just met his fate.

McClintock's team went to work preparing the boat, while Watson and Whitney handled provisioning and equipment. Without question the boat and its potential were welcome; General Beauregard was eagerly awaiting its arrival from the moment he was advised of its imminent departure and sent several messages to Mobile inquiring as to its whereabouts.

What is interesting here is that it becomes clear very quickly that the boat was under the operational control of the Confederate *Army*, not the Confederate Navy. General Beauregard was at the top of the chain of command for the boat for its entire tour of duty and the CSN appears to have had little to do with its employment, although they did provide valuable support in both personnel and training. And as the boat was prepared for its first mission, the team got a pleasant surprise – Horace Hunley himself arrived in Charleston on or about the 20th, along with other members of the Singer team who had remained behind in Mobile.

The boat was towed out to "The Cove", a sheltered area behind Fort Moultrie and now the site of Fort Moultrie Visitors' Center. The water was fairly shallow there, but it was at least protected and gave them easy access to the harbor approaches. It was just a matter now of deciding the when the boat was ready for her first patrol. That day came sometime between the 14th and the 20th of August, just around dusk. A night approach was mandatory – first, the chances of it being seen were just too great. Secondly, Admiral Dahlgren had a pretty good idea she was coming.

There had been several intelligence reports forwarded to the US Navy and Dahlgren that indicated the Confederacy was preparing some kind of 'secret weapon'. There was the deserter's report from Mobile Bay, [34] for starters. In addition, deserters and refugees came across to the blockade line on a regular basis, and

---

[34] See Chapter 2, pg 24.

from all accounts most of them seem to have known about the boat. To a certain extent, that's understandable – after all, the boat was brought into town on two open rail cars and was moored downtown for at least a week or so, and in the process the team seems to have been rather happy to discuss the matter with anyone who showed up.

We forget that this was a different time, when if you were a 'friendly' citizen, it was expected that you would remain close-mouthed and discreet, regardless of what you saw or knew. Sadly, this too had changed, and the Confederate military personnel and civilians who made it across to the Federal lines or ships were more than happy to trade whatever they knew for food, clothing, or simple safety.

There was one other source of information that tends to be overlooked in the history books – escaped slaves, or 'contrabands'.[35] The 'peculiar institution', no matter how benevolently some individuals may have administered it, tended to reinforce an attitude that blacks were intellectually inferior and were simply not intelligent enough to be spies. At best that was a badly misplaced belief and at worst it led to some serious security breaches throughout the war. Mark Ragan points out that on at least one occasion, slaves were directly assigned to help refit the boat[36], and there was a steady hemorrhage of slaves through the lines throughout the war. These were intelligent, trusted servants who could be relied on to do the job right – and when they escaped, they could be relied on to give solid, detailed information to Federal officers.

Mind you, forewarned was not necessarily forearmed: Dahlgren (and for that matter, his superiors in Washington) didn't necessarily dismiss the threat out of hand, but they didn't issue any specific intelligence assessment[37] or warning. There is

---

[35] So called after Federal General Benjamin Butler's May 1861 decree labeling escaped slaves as 'contraband of war' who could be confiscated and held – neatly bypassing any worries about taking of private property from Rebels, a matter of some concern early in the war. McPherson, pg. 355.

[36] Ragan, pg. 108

[37] Although this expression didn't come into use until the latter part of the 20th century, I have decided to use it here because of the concept that it represents – MJK.

CSS H.L. Hunley

also the likelihood that Dahlgren was confusing the boat with another weapon, which will be examined later.

The Competition: The Intelligent Whale, the US Navy's attempt at entering the realm of submarine warfare. Large and complex, she was a disappointment that wasn't even completed until a year after the end of the Civil War. (Photo courtesy of the United States Naval Historical Center, Washington, DC)

In addition, at just about this same time, some private entrepreneurs in the North were preparing their own submarine boat. Three men with the charmingly Victorian names of Scovel Merriman, Augustus Price, and Cornelius Bushnell formed a partnership in New Jersey called - logically enough - the American Submarine Company. They created a big boat - nearly thirty feet long, weighing two tons and capable of carrying *thirteen* men, though she only needed a crew of six, and calling it the *Intelligent Whale*. Hand-cranked like McClintock's designs, *Whale* used a slightly more advanced method of submerging than the new Mobile boat, with closed compartments being flooded then blown for surfacing by compressed air.

While more sure a method of getting back on the surface than the new Mobile boat's hand pumps, it limited the *Whale* to surfacing exactly once - after that, she would be stuck on the surface. In terms of offense, *Whale* would use a method of destroying her targets that had even more potential for disaster than the new Mobile boat: a mine would be carried on her back,

The Long Patrol

then a crewman in a diving suit would exit the boat through a surprisingly functional set of wooden doors in the keel. He would drag the mine to the unsuspecting target, fasten it to its hull, and then scamper back to the *Whale*, from where the charge would be fired by lanyard. This would also limit the *Whale* to operations in very shallow water, whereas the new Mobile boat could, at least theoretically, operate in fairly deep water.

Unfortunately, the potential profits from such an infernal machine led to prolonged lawsuits between Messrs. Merriman, Price, and Bushnell that delayed the construction and final completion of the *Whale* until April of 1866, by which point the war she had been intended to fight had been over for a twelvemonth. And to add insult to injury, the Dictionary Of American Naval Fighting Ships gently points out that the American Submarine Company "encountered great difficulty...in getting a crew to man her for her first test in Newark Bay."[38]

Her first test was, amazingly enough, a smashing success within its limits. Led by a misplaced former Army general named Thomas William Sweeney, *Whale* ventured out into Newark Bay to show her stuff. She did not disappoint. Sweeney personally donned a diving suit and bolted the mine to a target ship, then went back to the *Whale* and pulled the lanyard. The target went to the bottom, albeit in only sixteen feet of water.

But as any submariner will tell you, a kill is a kill, and the American Submarine Company began to rub its hands in glee at the thought of the piles of money they expected the United States Navy to start shoveling in its direction.

Alas, though engineers may propose, judges dispose. Courtroom fighting continued, and the *Whale* was eventually taken from its builders and awarded to a gentleman named 'Abe' Halstead. Mr. Halstead in turn made a deal with the Navy Department in 1869 for $50,000 dollars - contingent, however, upon *Whale's* successful trials under the aegis if the US Navy.

---

[38] *Dictionary Of American Fighting Ships*, 'Intelligent Whale', www.history.navy.mil/danfs/index.html

It was not to be. It appears that the *Whale*, now at the Brooklyn Navy Yard, pretty much sat at dockside for three years until the Navy finally got around to testing her. With little if any maintenance being performed in the meantime, she very nearly followed the example of her Confederate cousin. Upon submerging, the *Whale* immediately began shipping water through poor or deteriorated packing. Her crew decided that discretion was the better part of valor and tried to surface, only to come up under a crane that was there to help with just such an emergency. It took some time to get things sorted out, but they did finally pull the *Whale* clear of the murky waters of Wallabout Bay. There had been considerable flooding, but her crew - which seems to have included Mr. Halstead, who hopefully had time to reflect upon the economies he had taken with his boat - was alive and well. The US Navy, while grateful it didn't have a boat full of dead civilians on its hands, was unwilling to go any further and cancelled its contract with Mr. Halstead.

The *Whale* was abandoned and ended up on display at Brooklyn for many years, then was hauled to the Washington Navy Yard before being restored and put on display at her current home in New Jersey. Some more-or-less contemporary accounts give her a body count among her own crew almost as grim as the *Hunley* ever had, but this seems to be the result of sensation-seeking writers and badly garbled accounts of that single exciting dive in Brooklyn one fall day in 1872. The Navy, on the other hand, got itself out of the submarine business for twenty-eight more years before the brilliant John Holland convinced them to try again.

In any event, James McClintock took the new boat – apparently referred to at this point as the *Porpoise*[39] – out for the first time in mid-August. And from the beginning it seemed as if the first Mobile boat, the *American Diver/Pioneer II* – might have risen from its silty grave and come back to haunt them.

First, the distances involved were greater than anything the crews had tried to travel before. From the Cove out to the

---

[39] Ragan, pg. 52

blockade line would have been at least six and a half nautical miles, a round trip of thirteen nautical miles – and that didn't take into account the maneuvering necessary to get into attack position, which turned out to be far more than had been thought. At most, the test runs in Mobile were perhaps a nautical mile or two and back, and from the Theater Street docks they had the current of the Mobile River to help them.

As far as getting lined up for an attack was concerned McClintock learned that there was a significant difference between a worn-out, unmanned and unarmed hulk moored in the mouth of some placid river and a fully capable warship on patrol off an enemy city. The crew was exhausting itself simply trying to get within range, much less set up an attack on a heavily armed, maneuvering target.

Although it appears that on at least some occasions early on a small steam launch was used to tow the *Porpoise* out, it's clear that if a ship of any kind was spotted moving out from the harbor, it would have given the US Navy a warning that something was up.

Second, the torpedo was – to put it gently – a problem. The torpedo was mounted on a board and towed roughly one hundred and fifty feet behind the boat. In the Mobile Bay tests, the torpedo was a bit unpredictable, but it was felt at the time that the crew could work within those limitations.[40]

However, once they were in Charleston Harbor, with drastically different winds and currents than Mobile had, the torpedo almost seemed to take on a life of its own. No matter what heading the boat took, it would go port, it would go starboard, and quite frequently even once the boat had stopped the torpedo would keep skimming happily along the surface…and then stop disturbingly close to the boat, which surfaced in the blind, unable to see what might be hovering above it.

---

[40] See Chap. 2, pg. 10

CSS H.L. Hunley

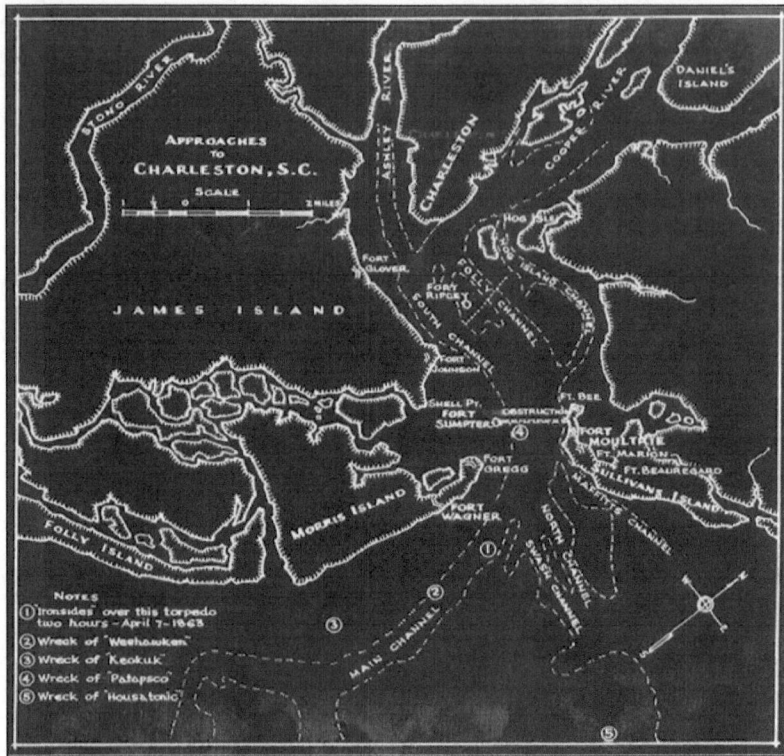

Battlefield Charleston: *US Navy map of the Charleston Harbor area and its approaches. Location #5 at lower right indicates the wreck of USS Housatonic. The Swamp Angel's emplacement was on Morris Island, while Fort Johnson was on the peninsula WNW of Fort Sumter. (Photo courtesy of the United States Naval Historical Center, Washington, DC.)*

Third, the boat's ability to stay running submerged seems to have been badly overestimated. Cranking for all they were worth used up the internal air supply (or at least made it unusable) within about twenty to thirty minutes. This, by the way, would seem to indicate that McClintock was taking her down at the earliest opportunity and keeping her there rather than running on the surface while still in the harbor and then submerging. The snorkel apparently did not work well enough to keep the crew in fresh air, and even deploying it for what good it could do meant sticking something above the water's surface that could be spotted by a lookout.

In the end, what all that meant was surfacing, popping both hatches, and standing motionless for at least a few minutes while fresh air recirculated, and then back down again. Bad enough while simply moving out into the channel, but it was positively suicidal while making an attack approach. Even at night, they were far from invisible on the surface. A sharp-eyed lookout on a clear, moonlit night could see her at a distance. Several ships had calcium lights installed – huge spotlights which had a flame applied to a cylinder of lime behind a carefully ground lens. The result was a remarkably harsh and long-ranged light that enabled Federal lookouts to spot even fairly small objects – driftwood, Mister Singer's torpedoes, or a submarine torpedo boat.

One would hope that McClintock and crew were at least cognizant of these problems before the *Porpoise* made her first sortie that muggy August evening, though one cannot shake the bothersome thought that they realized that serious operational difficulties were hiding in their future, and simply chose to either ignore them or hope that they wouldn't be as serious as they appeared. It would be no reflection upon the honesty or integrity of these men if that were the case – far from it. Rather, they probably believed with all their hearts that whatever drawbacks their boat might have, they could deal with them. In the end, they couldn't, but that does not subtract from their courage and inventiveness.

The Swamp Angel began its devastating assault on Charleston early on the morning of August 22nd. If nothing else had put the fear of God into the good people of Charleston, the Swamp Angel's murderous roar did.

Surviving letters from those who rode out the bombardment speak of paralyzing fear, sudden, horrifying explosions, and the acrid smell of smoke and blood. There were legitimate military targets in Charleston – the rail yards, the docks, the runners tied up to them, and the defensive batteries that ringed the town – but there appears to have been little effort on the part of the Federal gunners to discriminate between the guns and the stately homes of Rainbow Row, or between the docks and rail yards and the crowded homes and shops downtown. It was intended to terrorize the population, and no one made any effort to claim otherwise. The ships of the South Atlantic Blockading Squadron couldn't put the same amount of fire into the city, as it was much tougher for them to get within range, but they helped, and the days and nights of August 1863 began to run together in a numbing roll of unending thunder and terror.

And the *Porpoise* had yet to strike a blow for the battered people of Charleston. The exact number of patrols McClintock had performed by this point is unclear, probably no more than eight to ten. But there was now a disturbing feeling among the Confederate Army leadership on Sullivan's Island that James McClintock was not showing the aggressiveness that was needed in these trying times. The attitude seemed to be that one or two practice missions were fine, but the boat had proved its abilities, had it not? The military situation required – no, *demanded* – that an assault be made on the ships that cruised slowly offshore while turning the Holy City into charred wreckage. But every time McClintock went out he came back without results – no fear spread through the blockaders, no column of flame and debris marking the righteous vengeance of the Confederacy.

Exactly why? It appears to have been a combination of things, but they can be distilled down to a single point: the reality of taking the boat out against an enemy who was determined, angry, heavily armed, and ready to use overwhelming, unrestrained force. The rhetoric, the defiant attitude and speeches had run headlong into the United States Navy, and the Navy had prevailed.

We shouldn't be too hard on James McClintock. Human nature being what it is, it would have been surprising if he *hadn't* quailed a little in the face of the kind of force that was cruising off the South Carolina shore. And given a little more time, perhaps a few more training sorties, he may very well have found the confidence to go after one of the blockaders. He was *not* a coward, not by any stretch of the imagination – he was trying to invent an entirely new kind of warfare, and it may have been just a little bit beyond his abilities. Charleston however simply didn't have the time for him to learn and the Confederate States Army didn't have the patience.

The Long Patrol

What patience remained ran out on August 23rd. The US Navy had moved in close early that morning to sound reveille for Charleston, and for some time the ironclads and the shore batteries exchanged fire until a fog bank moved in and reduced visibility to near zero. There was certainly no sense in wasting ammunition, so they ceased fire and stood by. When the fog lifted, the Confederates must have thought that God himself was wearing Southern gray.

Less than a thousand yards away from Fort Moultrie – easily within range of the heavy cannon mounted there – the Federal ironclad frigate USS *New Ironsides* had run herself hard aground. Apparently the gunners at Moultrie had a hard time believing their good fortune, because they seem to have been a bit slow to get into action. *New Ironsides* had already proven herself to be a bit less resilient than her designers had hoped, and her handling was not at all good – a stray current, a moment of inattention to the soundings, and she was high and dry, her propeller frantically churning the Atlantic mud in an effort to get out of danger.

Had the Confederate gunners been a bit faster they could have literally blown *New Ironsides* out of the water, but before they could sight in on the stranded frigate the fog rolled right back in and she vanished from view once more, and within a short time had managed to free herself. The CSA commander at Moultrie, General Thomas Clingman, was angry enough as it was – the performance of his gunners had, frankly, been embarrassing and the dressing down they got later must have been impressive indeed.

But General Clingman saved his strongest anger for James McClintock and the crew of the *Porpoise*. The exact times and locations are unclear, but it appears that McClintock had the boat on patrol that evening, and was almost certainly within range to go after the *New Ironsides* – but didn't, and that was the last straw for General Clingman.

There are some very solid, reasonable explanations for this, and one may very well be that McClintock didn't know that *New Ironsides* was even close by. No radar, no periscopes, near zero visibility even when they were on the surface – it might have been a complete surprise to him when he discovered the ship had been aground at all. Even if he did know and had been able to find the frigate, an attack would have been almost physically impossible. The boat was still in its original weapons rig with the towed torpedo and if *New Ironsides* was aground, there certainly wasn't enough room under her keel for the boat to pass under her to make the attack. It's feasible that there may have been just enough room under *New Ironsides'* fantail for the boat to squeeze through, but that would have taken a degree of skill McClintock didn't have and a degree of precision that the boat wasn't capable of.

Be that as it may, General Clingman was in no mood to hear any excuses regardless of their validity. There were words exchanged that night between the General and McClintock, and the next morning a detail marched the short distance out of Moultrie's fortifications to the little pier on the Cove. There, they advised McClintock that the boat was now under the direct control of the Confederate States of America. The services of Captain James McClintock and company were no longer required – nor wanted.

Most of the team headed back to Mobile, with a couple of notable exceptions. Horace Hunley and Baxter Watson remained in Charleston while McClintock himself returned to Mobile some time later, leaving the story except for a James Bondian postscript a few years later. Gus Whitney stayed in Charleston forever – he died of pneumonia within a day or two of the seizure, the first member of the crew to be lost.[41]

Now of course, the Confederate leadership had to get the boat into action again and this meant getting a new crew trained. The Confederate Navy was, quite logically, placed in charge of the effort. Volunteers weren't hard to come by, so getting a crew wasn't the hard part. The boat's new commander, CSN Lt. Arthur John Payne, now had to get them capable of getting the boat out to sea, sink blockaders and come back in one piece.

---

[41] Ragan, pg. 64.

The Long Patrol

Payne, an officer who had seen a fair amount of action already and was now assigned to the ironclad CSS *Chicora*,[42] apparently jumped at the chance to take the boat out and started working his crew hard. Another member of the new CSN crew was Lt. Charles Hasker, another English immigrant who had joined the Southern cause. Hasker had been aboard CSS *Virginia* with Frank Buchanan, and then wound up reassigned to the *Chicora* with John Payne. The two men seem to have hit it off well, and when the opportunity came to transfer to the boat they ended up as the new command crew.

Training proceeded immediately, more to simply get a feel for things than anything else, and by August 29th Payne and his men had managed to get the hang of moving the boat from point A to point B, as well as a few short dives. There seems to have been an understanding at this point – now that the boat was under the operational and tactical control of the CSN – that there would have to be a period of solid, comprehensive training before the crew was able to safely take the boat out on a combat patrol, but that time was limited.

The 29th dawned clear, hot and muggy on Sullivan's Island as Payne marched his crew to the boat from Fort Moultrie and got them underway. It appears that Payne had planned to take them out on patrol either that night or soon afterwards and wanted to get a little more practice in. Accordingly, he got them underway in Charleston Harbor for most of the morning and then made for Fort Johnson, a small Revolutionary War-era post on James Island almost directly across the harbor from Moultrie.[43] Sometime that afternoon, Payne had her on the surface near the wharf at Fort Johnson with the hatches open. She was behind and outboard of the steamer CSS *Etowah*, with *Chicora* apparently either close aboard or tied up astern of *Etowah*. Another vessel was passing outboard of the boat, though how close is a matter of question.

---

[42] J. Thomas Scharf, *History of the Confederate States Navy*, pgs 565, 693, 695.

[43] Interestingly, Scharf states that the boat was actually three-quarters of a mile further east at Fort Sumter (pg. 760), though this seems to be very unlikely – risking the boat that far out in daylight makes little sense, and there are no Federal reports witnessing the salvage activities that followed. – MJK.

CSS H.L. Hunley

John Payne was in the forward hatch, and the boat was either moving ahead slow or was drifting while a handling crew on the *Etowah* was preparing to throw a line over the side to the boat so that they could tie up outboard. Payne was apparently standing on the Captain's seat in order to get himself far enough out of the hatch to grasp the line. Charles Hasker was in the first seat, and behind him (in unknown order) were four more men from the *Chicora* – Frank (or Frederick) Doyle, John Kelly, Michael Cane, and Nicholas Davis. Closest to the after hatch were two other crewmen [44] and CSN Lt. Charles Sprague, who was apparently in charge of maintaining and firing the torpedo. Sprague was definitely in the last seat, and was helping to propel the boat as well as serving as a weapons handler. Everything seemed at this point to be a completely routine dockside evolution on a sunny August afternoon.

What happened next is beyond any question: without any warning, the boat suddenly plunged beneath the brown waters of Charleston Harbor, leaving only a patch of foam, roiling bubbles, and a terrified, sputtering John Payne splashing in the water. The alarm quickly went up, reinforced by the appearance of Charles Sprague and the unknown crewman, followed a few moments later by Charles Hasker. Hasker, injured during his ascent when the forward hatch slammed shut on his leg as the boat went down, was rescued by an old friend from *Chicora* – Midshipman Daniel Lee, nephew of Marse Robert. Within a few minutes, Payne, Hasker, Sprague and the Unknown were safe aboard the nearby ships – but Doyle, Kelly, Cane, Davis, and one other were trapped in the boat, unable to struggle over the crankshaft to a hatch. They were dead within moments.

Why it happened is somewhat muddied, but easily enough determined. Scharf says that she was swamped by the ship that was passing them as they prepared to tie up to *Etowah*[45], but we

---

[44] Both Ragan (pg. 70) and Wills (pg. 19) identify one crewman as Absolum Williams from the ironclad CSS *Palmetto State*. The other is still unidentified. Burton (pg 231) says the boat was actually tied up at Fort Johnson, but gives no details – which suggests the swamped-by-a-passing ship scenario. Burton also claims Unknown status for the crewman from *Palmetto State*.

[45] Scharf, pg. 760.

have an eyewitness account from inside the boat itself - that of Charles Hasker[46]. His account is not complimentary to Lt. Payne.

Hasker stated that Payne was trying to get hold of the line that had been passed down from *Etowah* and gave the order for the crew to start moving ahead. They had just started cranking when Payne became tangled in the lines – and then stepped on the diving controls. The boat had always been responsive, and now was no different. She dove straight under, water cascading into the open hatches as she headed for the bottom forty feet away. Payne went straight up out of the forward hatch, while Sprague went through the aft hatch, followed by the Unknown. As mentioned previously, Hasker made it out through the forward hatch, but the hatch cover closed on his leg and it was only with a superhuman effort that he avoided passing over to Eternity within the boat's iron embrace. Hasker stated that he went all the way to the bottom with the boat before freeing himself and making his way up.

For about three days, no one was really sure what to do next. As is the military's wont, inquiries and interviews were made to try and determine what happened. And as with any incident of this kind, with literally dozens of witnesses and survivors' accounts, there were almost as many different versions of what happened.

In the end however, the final verdict was that it was operator error – i.e.; Lieutenant Arthur John Payne, Confederate States Navy, did inadvertently activate the diving fins, causing the submarine torpedo boat to dive and fill with water, resulting in the loss of five sailors. Said inadvertent activation is considered to be due to Lieutenant Payne's becoming fouled in a mooring line and his unfamiliarity with the position and operation of the controls within the submarine torpedo boat.

Lieutenant Payne was not held criminally responsible for the accident. Rather, it seems that the speed and haste with which the CSN wanted to boat put into action was considered a primary

---

[46] Both Ragan and Wills cite Hasker's interview with W.B. Fort, "First Submarine in the Confederate Navy," *Confederate Veteran* 22 (1914)

cause of the accident, with Payne guilty of no more than a tragic misstep. Indeed, when General Beauregard ordered the boat raised, he made sure Lieutenant Payne was kept in the loop. Given Beauregard's reputation, this does not at all appear consistent with a finding that Payne had been negligent or incompetent.

Two divers who had previously been employed in laying torpedo fields in Charleston Harbor were given the job of somehow getting the boat off the bottom of the harbor and ashore for refit and repairs. Over the space of a week or so the wreck was prepared for salvage, and sometime after September 8th the boat finally broke the surface again. She was quickly pumped dry – mostly, anyways – and towed the few feet over to the Fort Johnson wharf.

The scene inside the hull must have been awful. The boat had filled with water, and at least one hatch was still open enough to allow marine life access to the boat's interior. The five corpses inside were badly swollen, and between the effects of decomposition and the aquatic fauna they must have resembled something from a Bosch painting of Hell's inner circle. The bodies were removed – apparently with some difficulty – from the hull and were buried at the old Mariner's Cemetery in Charleston. The cemetery later became the site of the Citadel's football stadium, but the remains, well, remained. They were re-discovered in the late 1990s and now rest at Magnolia Cemetery.

In the meantime, the boat was now tied up back at Fort Johnson while the local commanders tried to figure out what to do next. It would take some time to get her back to mission-ready status, so attention could be given to finding another crew. Although Lieutenant Payne was still officially the *Porpoise's* CO, it does not appear that either he or Lieutenant Hasker were volunteering to go back out, so now a command crew had to be located as well.

During the third week of September, a courier brought General Beauregard what he must have considered the answer to his prayers – a letter from Horace Hunley himself.

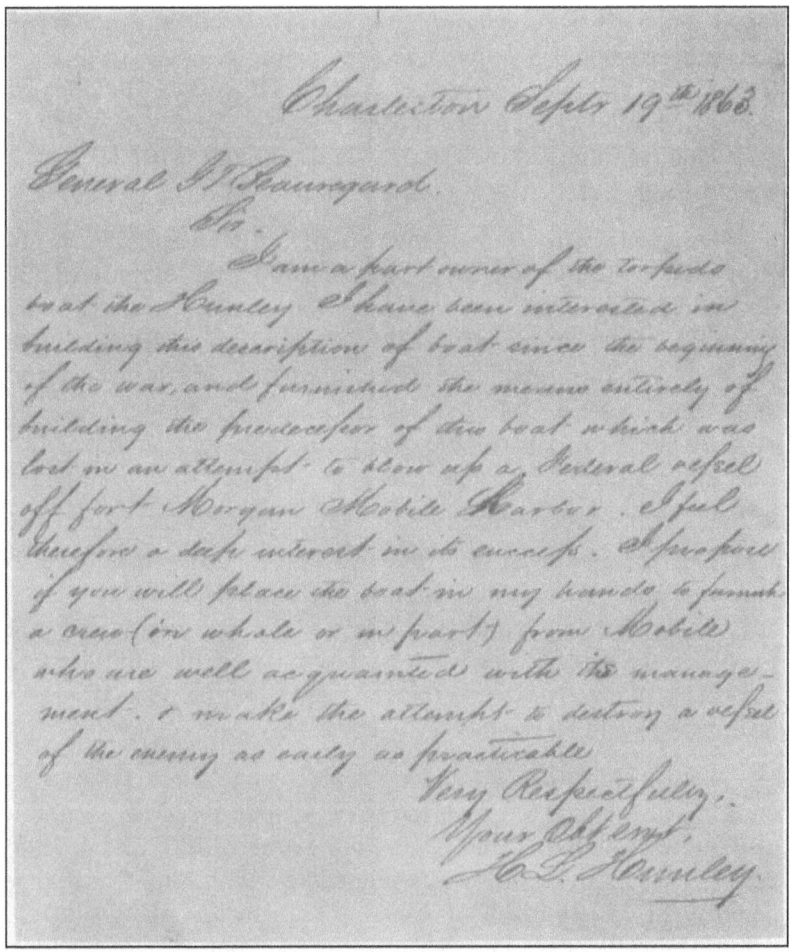

"...If you will place the boat in my hands...": Horace Hunley's letter to General P.G.T. Beauregard requesting that he be placed in command of the little black boat. (Photo courtesy of the United States Naval Historical Center, Washington, DC.)

    This was, as has been mentioned, a different time – today a civilian who wanted to contact a military commander in wartime would be passed off to a junior officer or NCO if they were acknowledged at all. But in those long-gone days, if a gentleman wrote to the local commander he could be reasonably sure that the officer in question would read it. And so it was that afternoon that General Beauregard read Hunley's proposal to repair and refit the *Porpoise*. Hunley was proper and polite but direct: he wanted to take personal charge of the boat, get it ready for combat, and take another crack at the US Navy. He knew of men in Mobile who could be counted on to form a new, well-trained and motivated crew that would take the war back to the enemy. Very respectfully, your obt. Svt., etc.

Beauregard probably knew who Hunley was. After all, Hunley was still (as far as we know) still in the employ of the Confederate government, working on ways to break the blockade, still a comparatively wealthy man with friends in high places. He had run the blockade on more than one occasion, and that was in a time and place where successful runners had the kind of adulation we reserve today for rock stars and athletes. Hunley was not only a gentleman, but also a brave Southerner whose devotion to the cause was beyond question.

With that in mind, Beauregard didn't need much time to make his call. A day or so later, he gave the order for Hunley – now with the honorific 'Captain' – to take charge of the *Porpoise* and prepare her for service, with a gentle reminder that she was still under military control. Anything he needed was to be approved and forwarded to Beauregard's staff. [47]

Horace Hunley had to have been ecstatic. Here at last was not only his chance to prove the basic soundness of his idea, but also quite frankly to find glory and perhaps still find a little profit. All of those things combined to make him thoroughly determined to get his boat back into the water and make his name. Hunley still had access to McClintock, Alexander, Watson, Dixon, and the rest of the team at Park and Lyons. He had the ability to get the boat refitted, and he knew he could train a new crew to take the *Porpoise* out and make it work.

There was, however, one possible problem that General Beauregard does not seem to be aware of: as nearly as the

---

[47] Interestingly, in his letter to Beauregard, Hunley himself refers to the boat as 'the *Hunley*.' As we know that at least some references have been made to the boat as the *Porpoise*, it seems reasonable to conclude that perhaps Horace Hunley had decided that since he had paid for the little black boat, he at least deserved the honor of having his name on her flank, officially or not - MJK

The Long Patrol

evidence can be read, Horace Hunley had not been in her much, and may never have spent a single moment inside the *Porpoise*.

It's probably a good idea to review some points – Hunley helped design and pay for *Pioneer*, but someone else commanded her on her trials. He helped design and build *American Diver/Pioneer II*, but there is no indication that he ever went out aboard her. The *Porpoise* was again the beneficiary of his ideas and pocketbook, but the evidence suggests that he never saw her in the water until after he arrived in Charleston on August 20th. It's possible he may have gone aboard her and even gone out on one of McClintock's training sorties between the 21st and when General Clingman took possession of her on the 24th – but he wouldn't have been in command.

The net result would be like taking the engineers who design and build modern attack submarines, along with the legislators who fund them, and putting *them* in charge of the boat. Although he didn't know it, General Beauregard was giving a great deal of control of the most advanced submarine yet built to someone who knew little or nothing about operating it.

Captain Hunley jumped into action, sending for men from Park and Lyons to report to Charleston as soon as possible. George Dixon was in this group, as well as Thomas Park, son of one of the machine shop's owners. William Alexander was to have gone, but Park was able to convince him to stay so he could go instead. Robert Brockbank, Charles McHugh, John Marshall, Henry Beard, and Joseph Patterson filled out the rest of the team.[48] Only one member of the old McClintock crew returned - torpedo specialist Charles Sprague, who had managed to escape the boat's sinking on August 29th.

Once the Alabamans arrived, everyone was put to work refitting the *Porpoise* and, just as importantly, getting her cleaned out. By the beginning of October, *Porpoise* was back in shape and ready for sea again. It is at this point that George Dixon assumed command of the boat – he had not only the experience in her design and construction, but he had been

---

[48] Wills, pg. 19.

aboard her during the Mobile trials. He was the best choice of the men now assigned to her to command her. Hunley seems to have organizing the entire training and refit operation with no direct command responsibilities.

Now operating from Fort Johnson, *Porpoise* went out daily and on some occasions nightly to make practice attacks out in the Cooper River, about two miles north. Her target of choice was the receiving ship CSS *Indian Chief*, more or less permanently moored in the river's channel. The boat was still rigged with the towed torpedo at this point, and to simulate it, Hunley and Dixon had secured a small barrel to the end of the firing line and would execute their practice attacks by making high-speed runs at the *Indian Chief*, then diving beneath it and then surfacing on the other side once the barrel had impacted the hull of the ship.

Lt George Dixon CSN. Source U.S. Navy Historical Center

Dixon – a tough, dedicated officer with combat experience – had them out there running attack after attack after attack, in clear view of the men who were defending Charleston. Once again, the boat's capabilities were proven – if she was able to get to a target, she was able to strike it. By the second week of October, Dixon had the men solidly drilled and working together as an experienced and knowledgeable team.

October 15th was dark and cloudy, perhaps the result of a near miss from a late season hurricane that was scudding by out in the Atlantic. Dixon had another day of training scheduled, but for reasons unknown he was not aboard the *Porpoise* – most likely, he was ashore working on getting supplies and provisions ready, or

The Long Patrol

setting up the boat's return to Sullivan's Island prior to commencing combat patrols again. In any event, Dixon wasn't at Fort Johnson that morning, an absence that would ultimately change history. He would not be the one to stand in the boat's forward hatch to take her out.

That honor belonged to Horace Hunley.

Whether or not Hunley had Dixon's approval to be at the helm is unknown. CSN Lieutenant Charles Sprague, who was the second ranking military member aboard, would have been the logical choice to take command of the boat in Dixon's absence. On the other hand, Sprague was a weapons specialist and he may not have had the training in the boat's operation and control to take her out. It does seem reasonable though that no one had any objections to Hunley taking command on October 15th – with the exception of Sprague, everyone else knew Hunley and was aware of the extent of his knowledge about the boat. If they had any concerns or objections, no one raised them, either aboard the *Porpoise* or with higher authority.

Hunley got the *Porpoise* underway at 0925, moving her easily out into the channel. It was raining and visibility was short, but on the Cooper the boat was easily visible, from both the shore and the deck of the *Indian Chief*. Hunley brought her about, lining up on the *Chief's* weathered hull before crouching down into the hull and securing the forward hatch. After a brief pause, *Porpoise* began to move steadily forward, gaining speed with each turn of the propeller, the practice torpedo scooting along behind it. Once the boat had gotten up to speed, Hunley gave the order to take her down at 0935, throwing the ballast valves wide open. *Porpoise's* wake began to climb up over the bow, past and over the forward hatch, and then the aft hatch. Another roll of the wake and she was under, the only sign that there was anything moving at all was the torpedo, now racing along seemingly of its own volition atop a V of water streaming back towards the dock.

There was a resounding THUD as the hollow torpedo slammed against the side of the *Chief*, followed by a rumble as it bumped down the hull - most of her crew barely even noticing, so commonplace had the exercise become. A few crewmembers, idling on deck, turned casually to the other side of the ship to watch the *Porpoise* slip back up onto the rain-stippled surface of the Cooper River. They watched, looking at the spot about fifty yards away where the boat would surface, then coast to a stop while the crew opened the hatch, evaluated the run, then turned around to do it again. They waited. Fifteen seconds…thirty…a minute…nothing. Other crewmembers, casually expecting to see the *Porpoise* appear, suddenly realized that it *hadn't*. They started walking, then running to the rail, and only then did they see the thin trail of silver bubbles winding up from the brown water.

There really was nothing they could do. There was no lack of men willing to go over the side to try and rescue their comrades, but there was no way to even locate the *Porpoise* in almost sixty feet of silty, opaque river water. And by the time they had found them, everyone aboard would have been dead. There was no more than an hour or two of air in the hull even under the best of conditions, and even less if she was partially or mostly flooded. Submarine Torpedo Boat *Porpoise* was down, its crew lost and presumed dead. In just over two months of operation, it had succeeded in killing thirteen men, not one of them a Federal sailor.

As the shock wore off, the divers went down once again and got a good look at the boat. She was nose first into the mud, her stern pointing upward at a fairly acute angle. The diving planes were full down, and most mysterious of all, the ballast weights – whose release would have sent the *Porpoise* bobbing back up like a cork – were still firmly attached to their place on the boat's keel.

While preparations were made to bring the boat up once more – as much to give the crew a decent burial as anything else – General Beauregard was having some very strong second thoughts about any further use of the boat for any reason. The sole surviving member of her second crew, Lieutenant George Dixon, could only stand on the dock in the rain and watch the slow, tedious work needed to get the boat and his friends back to the world of light and air.

The Long Patrol

The submarine torpedo boat was gone for a second time in less than two months. It would take time to find out why – and one could count on General Beauregard finding out why – but the undeniable fact was that one accident could be overlooked and explained. Two indicated something very, very wrong. Regardless of where the blame lay, regardless of what her potential was, the boat was now considered a jinx, a hoodoo ship. She had not traversed the depths protectively at all, indeed the exact opposite – she'd killed more of her own than the enemy. If Beauregard had his way, there would be no more trials, no more practice sorties. Take her apart and put the iron to good use.

But for all Beauregard's determination not to let the boat endanger any more Confederate warriors, he would eventually give in and permit one more try – an effort that would make history, change the face of warfare, and in the process place itself at the center of one of the sea's most enduring mysteries.

"...the Peripatetic Coffin..." Confederate States Ship H.L. Hunley. For more than a century this drawing by R.G. Skerret and several similar ones were dismissed as being a somewhat fanciful recreation of the Hunley, but after her salvage in 2000, it was discovered that these drawings were indeed an accurate portrayal. (Photo courtesy of the United States Naval Historical Center, Washington, DC.)

V: "...O HEAR US WHEN WE PRAY, AND KEEP..."

There never was any formal investigation into the second accident, at least not in the sense that we know it. An accident aboard or loss of a modern nuclear submarine is a matter of national security and safety, and the United States Navy has elevated submarine safety and design to an art. No other nation on Earth has operated as many nuclear powered vessels for so long with as little incident or injury. In 1863, however, there was no formal method to investigate such a thing. It was simply a matter of taking what you knew and applying it the best way you could. That effectively left General Beauregard with the educated guesses of Lieutenant Dixon.

We don't know what was said at that point, but they would have certainly discussed the matter to try and come to some understanding of why the *Porpoise* dove headlong into the mud of the Cooper River. Dixon never left any writings giving his opinions, but two other men, both intimately familiar with the boat did – Alexander and James McClintock, and both men were pretty much unanimous in their opinions. While they examined the facts that they knew, the divers were in the process of bringing the *Porpoise* back up, which they did sometime during the first week of November 1863. There had to have been great hesitation to open the hatches, especially as hollow, echoing thuds whenever parts of the hull were struck indicated that it wasn't filled with water, and that meant that the men inside hadn't died fast, relatively painless deaths.

So, once more, the boat was dragged to dockside manned only by the dead. Someone found a wrench and unbolted the hatches from the outside, prying them up and off with a crowbar. As they came off, there was a nauseating gust of air from inside as decomposition gases – and God alone knew what else – burst out under pressure.

Horace Hunley was still at his post forward, his body locked

The Long Patrol

by rigor mortis into a position that indicated he'd been trying to get the hatch open – an impossibility given the water pressure nine fathoms down. He was also clutching an unlit candle – a detail that will have significance later. Behind him, the forward ballast tank valve was in the full 'open' position, but its control handle – not so much a handle as a wrench that fit over the valve stem – was at the bottom of the boat.

Piled atop the handle were the bodies of Robert Brockbank, Charles McHugh, John Marshall, Henry Beard, Joseph Patterson, and Lieutenant Charles Sprague, who had managed to escape the *Porpoise's* iron grip on August 29th. In the aft hatch was Tom Parks, whose father's ironworkers had unknowingly built his coffin. He too was found apparently trying to open the hatch against the unyielding pressure of the Cooper River. Apparently the condition of the bodies was as bad – if not worse – than after the accident on August 29th; there are indications that this time the crew had to be at least partially dismembered in order to get them out.

In his book, Mark Ragan synthesizes the available data and puts together a logical explanation that explains exactly what happened.[49] Put simply, Hunley gave the order to dive as they approached the *Indian Chief*. Upon doing so, and with the boat at full speed, Hunley assumed that the boat needed more ballast, apparently thinking that they were diving too slowly – so he moved the ballast valve to the full 'open' position. The boat was at this point either nearly ballasted for diving or fully so, but either way the result was to send her charging straight under.

In and of itself, this wasn't fatal – all Hunley would have had to do was close the valve, abort the dive, pump out the forward tank, and try again. But when Hunley buttoned up the boat before taking her down, he had forgotten to light the candle that was their only source of illumination once they were underwater. At this point the boat is now several feet under the Cooper, in darkness – and instead of immediately terminating the dive, Hunley decided to try and light the candle instead.

---

[49] Ragan, Pgs. 97-98

CSS H.L. Hunley

Either not realizing or not remembering that the ballast tank was now taking on water at an alarming rate, Hunley was fumbling with the unlit candle when the *Porpoise* drove directly into the muddy river bottom at about four knots. The impact would have thrown Hunley against the forward bulkhead hard – but worse, it knocked the valve control handle off and to the bottom of the hull.

The boat is now in Stygian darkness, with water pouring over the forward bulkhead. Hunley is trying to find the handle while the men at the propeller crank are desperately trying to keep their heads above the rising water. At least one of them managed to get a wrench on the bolt that attaches the iron ballast weights to the bottom of the hull. If he can get it turned, the weights will fall away and the *Porpoise* will bob to the surface like a cork, giving the crew some chance to get out. But either they ran out of air – based on what Tom Parks was about to do, a strong possibility – or in their panic simply didn't push hard enough. The bolt is found only partially turned, just a quarter turn or so away from success.

While this is happening, Tom Parks, unsure what has happened and with no way to communicate with Hunley, does exactly the wrong thing: he frantically pumps out the *aft* ballast tank. The stern then rises up to that odd angle the divers found her at – but the rising water fills the hull up to the bottoms of the two hatches, drowning the men at the propeller crank. A small amount of air remains in each hatch, but not much – Hunley and Parks survive their crewmates by not more than a minute or two, and by 0945 they were all gone. The Cooper River smothered whatever final panicked cries or murmured prayers there may have been.

Horace Hunley could, up to that morning, be justifiably called one of the luckiest men in the world. Politician, lawyer, inventor, planter, blockade-runner – but when his luck finally did run out, it was as if the Gods had decided to make up for all the favors they had granted him before. An unlit candle, a moment's inattention, a loose valve fitting – in fiction, that would be called two coincidences too many, and a reader couldn't be faulted for slamming the book closed with a snort of disbelief and replacing it on the shelf. In reality, it put eight coffins on the long ride to Magnolia Cemetery.

Whatever his other faults, Pierre Gustave Toutant Beauregard knew how to lay on a funeral. He ordered a full-dress military ceremony for as soon as Hunley's remains could be recovered, which happened on Saturday November 7th. Two full companies of infantry and a military band – more than two hundred men, in total – escorted Hunley to his rest on Sunday the 8th. They marched through the gate off Meeting Street and filed in dignified procession to the far eastern side of Magnolia's moss-draped silence. There, in a quiet Episcopal ceremony, Horace Lawson Hunley was laid to rest beneath a glowing white marble stone that read:

CAPT H. L. HUNLEY

Aged 39 Years

Of Louisiana

A native of Tennessee

Who lost his life in the

Service of his Country

On the 15th of Oct 1863

The rifles of the Honor Guard would have spoken in what Douglas MacArthur once poetically called their 'mournful mutter', and the women in their billowing black mourning garb would have wept for him before they too turned and left for home.

One hopes that no one commented on the irony of the fact that Horace Hunley will sleep forever overlooking the river that claimed him. The rest of the crew was buried beside Captain Hunley on Monday the 9th, and it appeared that any further hope of getting the *Porpoise* into action was buried with them.

CSS H.L. Hunley

"...A look of confidence and assurance..." A memorial plaque to Horace L. Hunley at the Submarine Library, Groton, CT. (Photo courtesy of the United States Naval Historical Center, Washington, DC.)

General Beauregard was furious, and there's little doubt that he let Lieutenant Dixon know it. Thirteen dead Southrons – one of whom was a highly connected Government employee – quantities of scarce supplies and support provided, and the only tangible result was the financial enrichment of the salvors and the undertakers. Beauregard was solidly convinced that whatever *potential* the boat may have had, the risks of operating it far outweighed it. Enough was enough.

Dixon had already been back to Mobile in person to break the news to his old friends, a three day trip that would have given him plenty of time for reflection. When he returned to Charleston a few days after the funeral, he'd brought Will Alexander back with him. On the long ride back from the Gulf, the two officers had probably spent most of the trip trying to figure out exactly what had gone wrong and once they'd done so to try and put together one final pitch to General Beauregard. It must have been an impressive one, for Beauregard sent out word that evening that Dixon was to take charge of the boat and get her ready for action one more time. There was a caveat to Beauregard's permission – the boat could only be used on the surface. They were most emphatically *not* permitted to dive.

Obviously, this was a serious restriction on a submarine, but since Lieutenant Dixon was a Good Soldier he saluted smartly and went to work. He probably did so with his fingers crossed, however, because the boat was seen diving and surfacing many times over the next few months. For his part, Beauregard does not seem to have inquired too closely as to whether or not his orders were being followed – Will Alexander's later writings state that they couldn't even get a member of his staff to come out and observe an endurance test, so there was probably not a great deal of supervision from CSA Charleston.

In fairness, Beauregard had other things to worry about, since the US Navy was now hammering the city to one extent or another every day, and the possibility of a direct assault on the city could never be completely ruled out. And since Beauregard himself had a somewhat flexible concept of the word 'orders', it's likely that he simply decided that he had other priorities and what he didn't see couldn't hurt him.

By mid-November, Dixon had the supplies he needed to get the boat back into shape. He also requisitioned and got ten slaves, who probably had the unpleasant task of cleaning out the interior. Dixon and Alexander meanwhile literally tore the boat apart as far as they could, checking every rivet, every screw, every gauge and every stuffing box. Given the boat's history up to this point – four months in service and almost a month of that submerged and flooded – it was only prudent.

I believe, however, that these stuffing boxes – the points at which the control rods and the propeller shaft passed through the hull – are a key to later events. Keep in mind, however, that this had to be done on an *ad hoc* basis as none of the men who had built and designed the boat were available to assist. If any drawings accompanied the boat from Mobile they would have been used, but although these men were intelligent and capable, they weren't trained engineers. The possibility that something went wrong during the boat's reassembly cannot be completely discounted, though it is remote: the boat did make several safe and successful patrols before her loss, without report of incident.

Once the repair/refit was completed, Dixon and Alexander had an even greater challenge ahead of them: finding a new crew. There was no lack of enthusiasm in the lower ranks, but the commissioned members of Charleston's defenders, both CSA and CSN, were less than enthusiastic. On top of that, General Beauregard was suddenly wavering again about letting them go back out. Since Beauregard was never known as a ditherer, it's possible that his staff and some of the local naval staff expressed concerns and objections to him and he felt he had a duty to at least listen. Fortunately, Dixon and Alexander had picked up some of Hunley's sales ability for they were able to convince Beauregard to stay with his original call. Beauregard did put two more conditions on things though – first, they needed to clear the whole recruiting effort through Commodore Duncan Ingraham, the CSN commander at Charleston. Secondly – and Beauregard was adamant on this account – they had to fully brief every potential crewmember on the boat's unpleasant history.

A quick trip to Ingraham's HQ got the two officers an interview with the Commodore, who listened politely, if a bit skeptically, and gave his blessing. Ingraham directed them to report over to the CSS *Indian Chief* – the ship *Hunley* had been running mock attacks against on October 15th – to ask for

The Long Patrol

volunteers. The *Chief* was a receiving ship – that is, newly assigned and recruited sailors were billeted aboard her until they could be sent somewhere else, as well as being the depot ship for torpedo operations in Charleston Harbor. The men serving aboard her would have at least understood the theory of how *Hunley* and her weapon were supposed to work, and if anyone was likely to listen, it would be them. Dixon and Alexander headed out to the *Chief* and paid their respects to the skipper, who was probably most doubtful as to whether he'd ever see any of them again. But afterwards, he dutifully led them up on deck and mustered his crew.

Photographs of enlisted personnel in the Confederate Navy are hard to find. These Confederate sailors are probably typical in appearance. Source: U.S. Navy Historical Center

We are so used to the sharp, professional appearance of even the most junior Seaman Recruit in today's US Navy that we forget that as recently as the late 19th century dress and appearance regulations were far from standardized – or for that matter, enforced. The situation in the CSN was worsened by the fact that there never were enough uniforms to go around. That meant that the vast majority of enlisted sailors made do with whatever they brought from home or what they could beg, borrow, or in some cases steal. Since shoes tended to deteriorate and fall apart rapidly at sea – and were expensive for the average sailor – many sailors went barefoot. And since fresh water was always scarce, shaving often was too. That resulted in luxuriant beards and mustaches that never would have passed muster ashore. The frequency of bathing and clothes washing was also low, with results that the reader can imagine.

So when Lieutenants Dixon and Alexander viewed the pride of the Confederate States Navy lined up before them, they must have thought that they had somehow stumbled upon the last

CSS H.L. Hunley

stronghold of the Barbary Pirates. However, having gone this far, there was no turning back.

Dixon made his speech. He had seen combat – and been partially crippled in the process – so like most veterans, he probably kept the 'glory and honor' to a minimum even in an era known for flowery language. He would have told them that he needed strong, solid men for hazardous duty. That would have been no surprise to the men aboard the *Chief*, as most of them had seen the boat go down on October 15th.

Dixon would have told them that it would be tough, dangerous, and quite possibly thankless – and it might just turn the tide. Short of a great many unlikely things happening, nothing could by that point. But these men didn't know that and if they did, they didn't care. They'd made their choice and here was their chance to show they were worthy of the trust of a new nation.

Exactly how many men were even there is unclear, but five of them ended up going ashore with Dixon and Alexander that afternoon. First and most senior was Bosun's Mate James Wicks, CSN – the first Chief of the Boat. Next was Seaman Joseph Ridgeway, a Quartermaster from Richmond, Virginia. Seaman Arnold Becker, late of the great state of Louisiana, followed him aboard. He had been a cook, and by all accounts a good one – but a young man who wanted to do more than just cook, though. Seaman Frank Collins, CSN, another Virginian, hometown unknown. Finally was Seaman C. Lumpkin or Lumpkins – little of his previous life is known at all, but we do know the manner in which he went to glory. The five men gathered their gear, said their goodbyes and headed ashore with the Lieutenants.

The new submarine base was in the little village of Mount Pleasant, across the Cooper from Charleston. Today, it's known as a quiet, high-end suburb that is also the home of Patriot's Point Naval Museum where USS *Yorktown* (CV-10) the legendary Fighting Lady, rests in honored retirement. But a hundred and forty-one years ago, Mount Pleasant was a fishing community with a few small farms and a dozen small creeks and

The Long Patrol

streams running through it to empty into Charleston Harbor. The boat was towed there the same day that Dixon got his new crew assembled, and the next day they got to work.

Dixon's first move was a smart one. He had the boat hauled out of the water and put up on railroad ties. He and Alexander then turned to in making the crew learn the boat inside and out, from design to operation to tactical employment. Remember, both Dixon and Alexander had 'seen the elephant' – both men would have understood the old adage that the more you sweat in training, the less you bleed in battle.

Reveille was at sunrise every day, and as the holidays approached, it got progressively colder in South Carolina at night. The men would have arisen from under their thin blankets, marveling at their breath clouding in the air before their faces, and gathered together some kind of breakfast – perhaps some eggs and grits purchased from a local merchant if possible, but more often nothing if not, followed by whatever passed for coffee by that point, usually chicory and at one point later on, acorns.

Then they'd march down to the boat – now referred to, but never officially christened, Confederate States Ship *H.L. Hunley*[50] – and get to work. Dixon used the beached *Hunley* like a simulator; running exercises over and over and over again until the crew could literally perform their actions with their eyes closed. All things considered, that was only wisdom. After about a week of nonstop, dawn-to-dusk training, Dixon decided they were ready for a spin about the harbor and the *Hunley* was dropped back into the water. She would not leave it again for one hundred and forty-seven years.

The first trip went smoothly as they did a quick out-and-back into the harbor, each man performing flawlessly. Dixon wasn't through training his men, not by a long shot, but that first trial run had to have brought a smile to the young skipper's face. The

---

[50] There does not seem to be a record of an 'official' christening ceremony, either as *Porpoise* or *Hunley*. More than likely the decision was made by Dixon and Alexander to honor their friend, and the name became common use among the crew and Charleston HQ personnel - MJK

CSS H.L. Hunley

only unnerving aspect of the trip was that they were able to look up through the deadlights and see the US Army and US Navy throwing rounds into Charleston. The bombardment wasn't as steady as it had been earlier, but it was still doing indiscriminate damage and keeping the citizens of Charleston in a perpetual state of fear.

Thirty days after he began, Dixon had his boat rebuilt and his crew trained sufficiently that he was ready to take the *Hunley* out on a combat patrol. Dixon and Alexander had planned the patrol with as much care and precision as they could, making sure they knew as precisely as possible the layout of Charleston Harbor, the currents that raced through and by it, and the treacherous sand bars that lay outside it. Dixon made sure the men had a little extra food, some water, and possibly some distilled spirit to take the chill off – after all, it was December. Someone even had the foresight to bring a chamber pot along for those awkward moments when nature would call.

It had been four months since the boat had been on a patrol, and there was a lot riding on their performance that evening as the crew squeezed through the hatches and took their positions. Other than Dixon being at the conn in the forward hatch, we don't know where everybody else was, though it's likely that Alexander was at the other end, and Wicks probably in the first seat, just aft of Dixon. At around 2000 hours, the *Hunley's* crew started turning the crank and they moved quietly away from dockside, the towed torpedo floating unsteadily astern.

The first night out was a mixed bag in terms of mission performance. *Hunley* was able to get out on her own past Fort Sumter and reached the blockade line, but only at the price of the utter and absolute exhaustion of her crew. From the new base at Mount Pleasant, it was a staggering *sixteen* nautical mile round trip, simply too far against strong winter currents and seas. That was easily enough remedied, though it would take time to locate and prepare a new base.

Forts Johnson and Sumter were now under regular artillery fire from Federal units on Morris Island, and the original base on the Cove behind Fort Moultrie soon would be. Some effort and thought was going to have to go into the search, but in the meantime it would be everything the crew could do to get *Hunley* in range of her prey, much less pull off an attack. The evidence indicates that General Beauregard understood this limitation and was willing to accept it, at least for a while.

Secondly, the matter of internal illumination was becoming a real problem. The only truly practical method was to use an ordinary candle, but that not only used precious oxygen but also didn't last long. At best, the candles would burn about ten to fifteen minutes while they were under way, at which point Dixon was operating the *Hunley* blind. He could still see out the forward hatch deadlights, but the compass and depth gauge were all but invisible and the rest of the crew was mostly in darkness, something that could not have helped morale. They were making about twenty minutes underwater before surfacing.

And finally, a problem that arose like a revenant from the past – the seemingly possessed towed torpedo. During the training runs inside the harbor, the torpedo had been on good behavior, and McClintock apparently never got the boat far enough out for the currents and wind to be a problem – but Dixon did, and they were. The torpedo was behaving like a berserk toy, going left, right, fore and aft apparently of its own free will – and with no way to scan the surface before coming up, Dixon always had to worry that he might broach directly under the contact-triggered torpedo, with predictable results.

For the time being though, there was no remedy for that and all they could do was tread as warily as possible around the copper-cased monster. What finally did the towed torpedo in was an incident just after New Year's Day. Until a new base could be prepared, *Hunley* was still operating out of Mount Pleasant, but now she had help getting out to her patrol area: the steam torpedo boat *David*.

The *Davids* were a series of more-or-less identical torpedo boats – small wooden darts powered by steam engines that made them wickedly fast for the time. They could be ballasted down low to the water, then use a spar torpedo – much like *Hunley's* but mounted on a long boom *ahead* of the boat – against Federal

CSS H.L. Hunley

ships. They were highly maneuverable and made *very* small targets. [51]

THE BEGINNINGS OF SUBMARINE WARFARE

A CONFEDERATE PHOTOGRAPH OF '64—THE FIRST "DAVID," FIGURING IN AN HEROIC EXPLOIT

"...Small, wickedly fast wooden darts carrying a spar torpedo..." David class torpedo boat aground at Charleston after the war. The bow is to the left of the photo. At least one David was brought back to the US Naval Academy at Annapolis, MD, and operated there for some years. This may be the David that attacked USS New Ironsides on October 5, 1863, but that cannot be confirmed. Source: U.S. Navy Historical Center

David's first skipper, CSN Lieutenant Commander William Glassell, (granduncle of a certain George S. Patton) scared the living daylights out of the Federals on October 5th by taking her out against everyone's favorite target, USS *New Ironsides*. The approach was silent, skillful, and utterly undetected until they were just a hundred and fifty feet away from the big ironclad, at which point her crew opened fire with small arms against the

---

[51] It should be noted that the last few *Davids* were finished with hand-crank rigs like the one that powered *Hunley* – it's not clear if any of these saw service, but at least a few of the steam powered ones survived the war, and in one case made it to Annapolis, surviving as an operating curiosity as late as the 1880s - MJK

The Long Patrol

oncoming attacker. Glassell's men, demonstrating remarkable composure as musket balls rained down around them, fired back and killed ENS C.W Howard aboard *Ironsides*. The torpedo impacted about three-quarters of the way back along the frigate's starboard side, strangely close to where the *Hunley's* torpedo would strike *Housatonic* four months later. The damage in sum wasn't fatal - a crack in her side, and a blown-out armory bulkhead - but it wasn't repairable on the line so the *Ironsides* would end up making for Philadelphia, albeit slowly, out of the war for almost a year.

However, an unexpected side effect of the torpedo's detonation ruined LCDR Glassell's triumph: the explosion threw a column of water fifty feet into the air. As the *David* turned hard a port to get away, she passed extremely close to the *Ironsides*, and the descending water came directly down *David's* stack, extinguishing her boiler. This could best be described as leaving her in an unfavorable tactical situation – dead in the water and still well within musket range of the now thoroughly angry sailors and Marines on *Ironsides'* deck. *David* drifted out of range, but Glassell was convinced the game was up and ordered his crew – Assistant Engineer James Tomb, Fireman James Stuart (who may have gone by the alias Sullivan[52]) and Pilot J. Walker Cannon – over the side.

The choppy waters quickly separated the crew, and a Federal supply ship picked up Glassell and Stuart/Sullivan. In the meantime, Tomb and Cannon were carried back towards the still-floating *David*. Taking a chance, they got back aboard and somehow managed to get the boiler relit, wheezing their way back to the Charleston waterfront. Tomb and Cannon were heroes (Tomb was promoted and given command of the *David*), while LCDR Glassell and Stuart/Sullivan spent a good part of the rest of the war as guests of the United States of America before being paroled. *C'est la guerre*, as General Beauregard might have said.

There would be one more unsuccessful attack by *Davids*, a fact that may have some bearing on later events, but for the most

---

[52] E. Milby Burton, *The Siege of Charleston 1861-1865*, pg. 220.

part they remained a far greater potential threat than an actual one. In and of itself, this was not an undesirable thing - having to watch for highly dangerous things coming out of Charleston as well as in tied up more Federal men and ships and might help a runner get in when it was needed - but on the whole they probably weren't worth the resources needed to build, man, and maintain them.

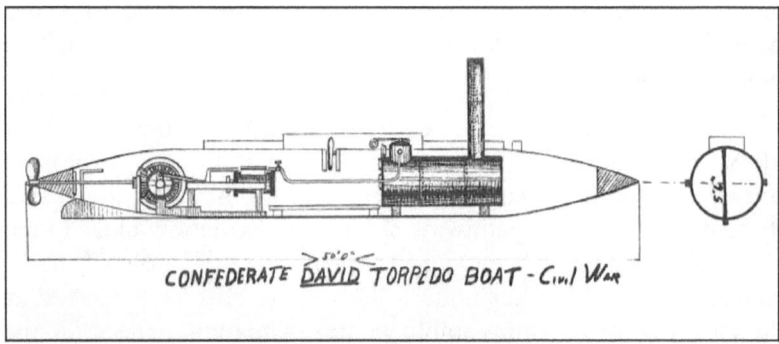

In the meantime, Tomb had been detailed to tow *Hunley* out of Mount Pleasant every evening and then cast her off near Fort Moultrie. This went fairly well until one night in mid-January when they got to the release point and the *David* came to a full stop in order to let the towrope go. *Hunley* slowed down as well – but the torpedo didn't. It continued to drift forward, *past* the *Hunley* and actually overtaking the *David*, not stopping until the torpedo line had fouled the *David's* rudder.

The two boats were now drifting together towards the mouth of the harbor, still tied together, and with a live, contact-detonated torpedo bumping gently against the *David's* fantail. All it would have taken was one solid impact and ninety-five pounds of black powder would have let go, tearing apart the torpedo boat's wooden hull and sending her straight to the bottom. Only the bravery of one of *David's* crew saved them – he went over the side into the cold, choppy water and got the torpedo line untangled. Tomb must have kicked up a roostertail getting out of the way and back to the Charleston waterfront.

The *Hunley* pressed on that evening, but after Tomb reported the festivities to his superiors the CSN commanders flatly refused to allow the *David* to act as a towboat again. Within a

The Long Patrol

week or two, *David* was given a partial ironclad hull and sent over into the Stono River to try her luck – but she was never able to even damage another Federal ship.

At roughly the same time, it was discovered that the Federal ironclads – the ships closest to the city and by default *Hunley's* primary targets – had deployed anti-torpedo measures. For the most part, they were chains or fenders hung away from the sides of the ironclads on booms and hanging down below the water's surface. These would snag and detonate a towed or spar torpedo before impact, drastically attenuating the effects of such a weapon.

Admiral Dahlgren had ordered this after a couple of very lucky breaks on his part. First, for reasons still unknown, LCDR Glassell had a fairly complete set of drawings and specifications for the *David* on him when he was captured. Secondly, two CSN sailors from the *Indian Chief* had stolen a small boat and made their way out to the blockade line to surrender. They turned out to have known quite a bit about the *David*, and a great deal about the *Hunley*.

All of that information made it to Admiral Dahlgren, who in turn finally took the whole thing very seriously. The threat of the spar torpedoes was bad enough. But now the South Atlantic Blockading Squadron was faced with what appeared to be an operational submarine torpedo boat. With the *Davids* there was at least a chance of spotting them – but with the submarine boat, Dahlgren had to acknowledge the possibility that he would never even know the boat was there until one of his warships was blown out of the water. However, even at this point, Dahlgren made a crucial error in assessing the threat:

"Flag Steamer Philadelphia,

Off Morris Island, S. C., Jan. 7, 1864.

"I have reliable information that the Rebels have two Torpedo Boats ready for service, which may be expected on the first night when the water is suitable for their movement. One of these is the David which attacked the Ironsides in October, the other is

similar to it.

There is also one of another kind, which is nearly submerged and can be entirely so; it is intended to go under the bottoms of vessels and there operate.

This is believed by my informant to be sure of well working, though from bad management it has hitherto met with accidents and was lying off Mount Pleasant two nights since..

"It is also advisable not to anchor in the deepest part of the channel; for by not leaving much space between the bottom of the vessel and the bottom of the channel, it will be impossible for the diving Torpedo to operate except on the sides, and there will be less difficulty in raising a vessel if sunk.

John A. Dahlgren,
Rear Admiral, Commanding
South Atlantic Blockade Squadron[53]*"*

In other words, Admiral Dahlgren told his commanders that they should operate in the shallows closer to shore – exactly what Dixon was going to be able to do with the spar-equipped *Hunley*.

All of this put together meant that the towed torpedo had drastically outlived its usefulness – but the truth be told, at that point no one was sorry to see it go. In its place a spar and tackle rig was designed, based on the rig that the *Davids* and a series of torpedo rowboats had already taken into combat. The rowboats, with a small spar torpedo attached. LCDR Glassell actually took one of these out (the exact date is uncertain) after the frigate USS *Powhatan*. Getting out to the blockade line in a rowboat had to have been as physically demanding and exhausting for them as it

---

[53] *Official Records Of The Union And Confederate Navies In The War Of The Rebellion*, Vol. XV, pgs. 227 - 238.

was for the *Hunley's* crew. In the event, the attack did not go at all well; Scharf states that one of the six rowers kept backing his oars, "from terror or treason", and Glassell had to abort the attack[54].

In any event, the spar torpedo would be mounted on the *Hunley's* bow and carried much like a medieval knight would carry his lance into combat. Most portraits and artist's impressions of the *Hunley* show it with a single spar, roughly fifteen to twenty feet long, bolted directly to the top of her prow, and the famous George Cook photo/painting shows her exactly like this. But another portrait exists, done by a gentleman named Conrad Wise Chapman on December 6th, 1863 while the *Hunley* was still ashore in Mount Pleasant – and it clearly shows a spar drooping downwards from a pivoting attachment at the base of the *Hunley's* prow, exactly where they found it when she was raised one hundred and forty-seven years later.

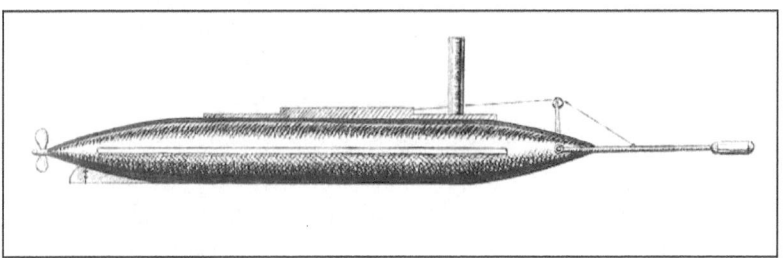

A David with its spar torpedo deployed. Source: U.S. Navy Historical Center.

The exact rigging of this spar is still a matter of discussion, because it will substantially effect the *Hunley*'s appearance – but we will pass that matter for now and ask the same question the sharp-eyed reader already has noted: we know the *Hunley* was still sailing with the towed torpedo in mid-January when they gave Chief Engineer Tomb and his crew heart palpitations that wild night off Fort Moultrie. So why is there a spar on her in early December?

I believe that the decision to go to a spar torpedo was

---

[54] Scharf, pg 754-755.

actually made in November, when the *Hunley* was pulled ashore for her repair/refit, and that Lieutenant Dixon was the driving force behind it. We know:

- The effects of the attack on the *New Ironsides* on October 5th. Although the true extent of *Ironsides'* damage was never known to the CSN, they knew from Chief Engineer Tomb's report that the torpedo *did* function as designed, and clearly well enough to knock one of the most heavily armored ships in the world clean off the blockade line.

- While training his new crew for their first mission, Dixon and Alexander were gathering all the information he could on currents, wind and sea conditions in his patrol area. He knew about the towed torpedo's wanderings, and since we know he was getting hard data on the sea and weather conditions they'd face out in the harbor, they *had* to have realized that the towed weapon would be a distinct problem – anything else would have been an unusual lapse on the part of two otherwise sharp and remarkably foresighted officers.

- *Hunley* was fairly maneuverable, but nowhere near sufficiently so to make any sudden changes in her attack approach as she got near a target. The towed torpedo would have swung wildly behind her if she needed to break off the approach, and there was a good chance that it would keep sailing on into the turning *Hunley*. That was a nonexistent risk with a spar torpedo. In addition, the towed torpedo would make a marvelous target for sharpshooters on the target ship – a spar torpedo would be invisible beneath the waves.

Three good, solid reasons to go with a spar mounted torpedo, and I believe they were taken into consideration and acted upon while *Hunley* was still ashore in Mount Pleasant, and of course it made far more sense to refit it while she was ashore than in the water.

Why then were they still risking patrols with the termagant torpedo? Dixon was probably testing different boom and rigging configurations for one thing – that would also explain two confirmed portraits of the *Hunley* with two different boom configurations. There is also the growing likelihood since *Hunley's* salvage that the upper spar wasn't intended for use as such but rather is actually a part of the rigging for the whole thing - therefore making the lower spar the primary mount.

Secondly, the torpedo was a bit different than the old towed one or the *David* weapons. It was larger (one hundred and thirty-five pounds of powder vice ninety-five pounds) and had been custom designed by Mr. Singer. Singer's torpedoes were made from thin copper or tin and fitted with contact detonators, as was the *Hunley's*, but this one had a massive iron or steel barb mounted at its forward end and was command detonated by line from inside the forward hatch.

A modern replica of the Hunley's torpedo, Picture courtesy Averasboro Battlefield Museum

The idea was that the *Hunley* would ram the torpedo into the side of the target. The barb would hold it in place as they backed away, slipping the torpedo off the spar. When they had backed off to about one hundred and fifty feet – hopefully far enough – the line would be pulled, a safety pin would fall out, and a spring-loaded firing pin would drop onto a primer rigged with two percussion caps and filled with rifle powder. Ignition – combustion - explosion. The entire rig was mounted on a pivot at the foot of *Hunley's* prow, and could be raised or lowered to suit the angle of attack or the mounting or removal of the torpedo.

I would also like to propose another possible problem with the Singer spar torpedo that does not seem to have appeared in previous works. Remember how the spar torpedoes were

CSS H.L. Hunley

intended to be used: slammed into the side of their targets at the end of spars being propelled at between 4.5 and eight miles an hour by a *Hunley* or a *David*. They were either wooden kegs with contact detonators or (in the case of the *Hunley* weapon) a copper or tin cylinder.

I respectfully suggest that the fairly light construction of the torpedo (Mark Ragan points out that the Singer spar torpedo had positive buoyancy and therefore would have been able to float on its own) would have meant that on impact the torpedo – pinned between the side of the ship and the heavy wooden or iron spar which in turn is being backed up by several tons of torpedo boat - may have deformed or been damaged to such a point that its structural integrity and the integrity of its explosive material and firing mechanism were severely compromised.

This could explain the failure of the *David's* torpedo to actually sink the *Ironsides*, and the failure of a later *David* attack on March 4th, 1864 when she made a perfect attack, but the torpedo failed to detonate.[55] I believe it also helps explain what happened to the *Hunley* on February 17th – but we shall return to that later.

There was one warning voice regarding the torpedo refit, and perhaps it should have carried greater weight – that of Chief Engineer James Tomb, the CO of the *David*. Almost a year after the loss of the *Hunley*, Tomb wrote the following:

"In my report to him (CSN Flag Officer John Tucker, Tomb's immediate superior) *I stated, "The only way to use a torpedo was on the same plan as the 'David' that is, a spar torpedo and to strike with his boat on the surface, the torpedo being lowered to eight feet. Should she attempt to use a torpedo as Lieutenant Dixon intended, by submerging the boat and striking from below, the level of the torpedo would be above his own boat, and as she had little buoyancy and no power, the*

---

[55] It should be noted that the *David* torpedoes were of different construction (wooden keg) and trigger assembly (acid cutting a safety wire, allowing the *David* to back off to a safe distance from the explosion). However, I believe the basic point - the strong possibility of moderate to catastrophic damage to the torpedo on impact - still applies - MJK.

The Long Patrol

chances were the suction caused by the water passing into the sinking ship would prevent her rising to the surface, besides the possibility of his own boat being disabled."[56]

Tomb also made one other comment as he ended his report, a sadly prescient one:

"She was a veritable coffin to this brave officer and his men."

A run of thoroughly unpleasant weather in early January had effectively suspended operations, so Dixon and Alexander took advantage of the delay to scout out a new base. It had to be within sight of the blockade line and close enough for *Hunley* to get out to it under her own power, yet sheltered enough that it offered some kind of cover. There had to be some kind of dock facility emplaced, and ideally some kind of military post close by. The best fit was on the northern end of Sullivan's Island, a few miles northeast of Fort Moultrie. The CSA had torn down a line of homes and farms to erect Battery Marshall, a coast defense battery that covered a small channel known locally as Breach Inlet. Commanded by CSA Lt. Col. O.M. Dantzler, it sat just a stone's throw up from the beach on the island's ocean side. A small dock sat on the Inlet itself, and the blockade line could be clearly seen – and engaged from - the Battery itself. Accordingly, General Beauregard ordered *Hunley* moved there during the third week of January, and a day or two afterwards they were headed back out on patrol.

The routine was a monotonous and exhausting one. They would arise around 1300 hours and form up for the seven-mile walk from their quarters in Mount Pleasant (no one ever seems to have considered moving them to Battery Marshall or even Fort Moultrie). They would arrive at around 1500 and check out the *Hunley*, making sure that she hadn't sprung any leaks or developed any other problems since they'd tied up that morning. Dixon insisted on daily training, even if they didn't go out that night, so it was out into the sheltered Back Bay area for at least two hours of practice. It was deep enough to submerge and run

---

[56] *Official Records*, Series II, Vol. I, pgs 334-335

CSS H.L. Hunley

mock attacks against some of the smaller ships that could moor there – but it should be noted that unlike the repeated live-fire tests with the towed torpedo, there don't seem to have ever been any similar tests with the spar weapon. It would have been easy enough to run simulated attacks with an empty spar, however.

There's one other thing that seems to be missing from the Breach Inlet deployment – any kind of maintenance other than routine pre-operational checks before departure every night. Mark Ragan was able to put together a paper trail that documents the life and movements of the *Hunley* and her crew on an almost daily basis – but after the November-December refit, there is no indication that the boat was pulled out of the water again before February 17th. Keep in mind that means that *Hunley* was in the water steadily from mid-December through to the middle of February – just a bit more than sixty days in cold, rough seas and including at least one or more groundings. Both the lack of a live-fire exercise and the lack of maintenance are two more links in a potential chain that could have led *Hunley* to the bottom of the Atlantic Ocean.

Once the *Hunley* has tied up after her training run at about 1730, the crew has around ninety minutes to eat dinner, relax a bit, and then mount the torpedo on the spar. In the meantime, Dixon and Alexander will head to the beach below Battery Marshall and try to select a target for the evening. Assuming they can, they will cast off at around 1900 hours. It will be a long, hard slog from there. Dixon will conn them out of Breach Inlet on the surface, then light the candle and take them down a few hundred yards offshore. From there it will be up and down approximately every twenty minutes until 2400.

Pushing the tiller over and bringing the *Hunley* about, the crew will start the long trip home. As has been mentioned previously, sometimes it's just too much – the winds and currents get the best of the crew, and all they can do is drift ashore, sometimes on Sullivan's Island, sometimes on the Isle of Palms on the other side of Breach Inlet, and wait for the pilot boat to come out and get them. Either way, it's 0530 or 0600 by the time the exhausted crew gets ashore, another half hour or so to secure the *Hunley,* then the two hour march back to Mount

The Long Patrol

Pleasant where they were hopefully asleep by 0830.

They didn't go out every night – Alexander says they only went out three or four times a week, and even then only when conditions were right [57] - and there were a number of reasons the missions could be scrubbed. The weather was probably the single biggest cause; they needed dark, moonless nights, preferably with fairly calm seas, and they didn't always get it. Alexander says they would talk to the Charleston pilots (the last time apparently on January 31st) to get a good idea of what the winds and seas would be like.

When the weather did cooperate, they had to pick out a target. The ironclads, stationed closest to shore, weren't specifically off limits as targets but Dixon was practical enough to know that the spar torpedo wasn't likely to take down one of the heavily armored monitors. The wooden-hulled ships were farther offshore, but usually one or two wandered in close enough to make it worth the effort. Unfortunately, the result was the same every time: as they got farther out the wind and currents might change, or the ship would hoist anchor and sail away, or the implacable clock would strike midnight and they would have no choice but to put about. Those had to be the tough ones; there were at least a few occasions where they were close enough to hear the voices of their target's crew, but simply couldn't catch up to them.

The days weren't much better. Once the crew had fallen exhausted in their cots, they only got about four and half to five hours of sleep – in theory. In reality, the long artillery duel between the United States Army, United States Navy, and the Confederate defenders of Charleston was still underway. Even seven miles away in relatively safe Mount Pleasant, the sound must have been like a continual, never-ending thunderstorm. They could stuff rags in their ears to deaden the sound, but they could never escape it entirely.

Their sleep must have been fitful at best. Admittedly one can get used to anything, but that does not mean their rest was

---

[57] Ragan, pg.150

peaceful. After a while – chilled and exhausted when they finally got to their quarters – the sleep they were getting could not have possibly been enough, and they would have started to suffer from sleep deprivation. There are reasons the United Nations classifies sleep deprivation as a torture, and after a month or so the crew of the *Hunley* would have understood them.

This leads us to another possible link in the chain – the crew's health. First, their diet wasn't what we would refer to as well rounded. Based on modern standards, they were – as a rule – seriously undernourished. They were probably only getting one meal a day, with perhaps some coffee in the morning before heading back to Mount Pleasant. We will return to this facet of the problem shortly. There also had to have been a problem with respiratory infections of one sort or another. Gus Whitney, may his soul rest in peace, died of pneumonia more than likely caused by climbing in and out of the cold, condensation-laden hull and into the warm Carolina mornings. The next two crews didn't survive long enough to develop any problems, but consider: Dixon and his men were going out to sea in late fall and early winter. The water temperature – had they ended up in it – would have killed them in about twenty minutes or so. The air in the hull would have been chilled as well, while living and working in close quarters guaranteed that whatever anyone caught would quickly make the rounds of the crew. That in turn seems to guarantee that most, if not all of the crew was suffering from a low-grade upper respiratory infection at one time or another.

All of this added together made for nine thoroughly unhealthy submariners. As a rule, the general health of Civil War combatants was not good on either side, but the Confederacy was hit harder – there was never enough food, never enough money to pay for it, never enough transport to get it to the troops. What did make it to the men was barely nutritious and doled out as sparingly as possible, which meant that they weren't even getting what they were supposed to. Had Dixon's men been assigned to less physically demanding and mentally duty ashore, it would have been difficult enough, but they could have held on as did tens of thousands of other

Confederates. But given what they had to do every night just to get to their boat, then crank the boat out to the line and back, then march home to perhaps three to four hours of fitful sleep...it was a recipe for complete physical and mental fatigue. And on the knife-edge they sailed every night, that was one link in the chain too many.

The Intended Victim: *The frigate USS Wabash, which had sat off Breach Inlet for some time before Hunley's final mission. She came off the line somewhere around February 10th-12th 1864 and was replaced by the frigate Housatonic. Wabash would survive until 1912, when she was burned off Eastport, ME, to salvage her fittings. (Photo courtesy of the United States Naval Historical Center, Washington, DC.)*

CSS H.L. Hunley

Out and back, out and back, over and over again, and with no result. From mid-January to February 17th, we can guess that they went out about eighteen to twenty times, and the results were the same every time – no attack, no sinking Federal ship to mark their success. They had a chosen prey – the big Federal frigate USS *Wabash*, which at twelve miles (Will Alexander's estimate) was the closest practical target. There was only one event that stood out in this period: an endurance test to determine just how long they could stay under.

It appears that Dixon intended to do something along these lines, but had not set a specific date to do so. Taking an evening when either the weather was too bad to go out or there were no ships close enough to make a run at, Dixon had the crew board and then took her into the Back Bay and headed for the bottom. The candle went out at about twenty minutes, and then the crew sat back, determined to hold out as long as possible - but no one was exactly sure how long that was until they got back topside. There was a brief moment of terror when the aft pump wouldn't function, but it turned out to simply be some vegetation that had been pulled inside it. Dixon looked at his watch when they opened the hatch: two hours and ten minutes. The TNT film *The Hunley* gave a superb and positively claustrophobic depiction of what the crew went through, and as with anything concerning the boat, the film can be strongly recommended as accurate and truthful.

There was one moment of grim humor – Dixon literally scared the wits out of the poor Confederate soldier who was on guard duty at the dock that night. He apparently didn't hear the boat approach from behind him and nearly jumped out of his skin when Dixon called to him to throw him a line. Once the Private's heart started again, he eagerly complied – then told Dixon that they were supposed to be dead. When they hadn't come back up after an hour or so, word had been sent to General Beauregard to let him know that the *Hunley* had killed its third crew.

Dixon sent his compliments to Beauregard as quickly as he could, notifying him that the reports of their deaths had been greatly exaggerated and that they were ready for service. It will

The Long Patrol

be noted at this point that the fiction of keeping the *Hunley* on the surface per Beauregard's orders was now in utter and absolute tatters, especially as CSA Charleston had actually been invited to witness the test and Dixon sent a message confirming their survival straight to Beauregard. One may assume that as the general situation had continued to deteriorate after Beauregard's initial directive, the general finally gave up and realized that the best - and only - way to use the little iron boat was to use her as designed.

On February 5th, the crew took a blow from which they may never have recovered: William Alexander was recalled to Mobile [58]. Alexander's engineering skills were impressive and in high demand as technological solutions rapidly became the Confederacy's last hope. In this case, it was a heavy breech-loading cannon that was being developed back at Park and Lyons. Dixon and Alexander had apparently been wanted back in Mobile for some time to work on one project or another, but this time someone there put their foot down and insisted on Alexander's return. Heartbroken about leaving the crew and not wanting to hurt morale by a drawn-out farewell, Alexander left that night for Mobile.

Dixon went out and collected three crewmembers to replace him, CSA Artillery Corporal C.F. Carlson and two gentlemen who have come down to us only as Miller and White. Like any men who live and work closely together as they perform a unique and hazardous mission, the crew had come to regard each other as more than just fellow crewmen – they *were* brothers, regardless of the differences in their ranks or stations in life, and they had come to function together reflexively, precisely, and supportively. Alexander, who probably sat under the aft hatch, had been an important part of the team – helping keep up morale and supporting the men who sat in the aftermost part of the *Hunley*, where the candlelight never quite reached. In turn, the crew knew that Alexander had the courage to sit there with them night after night, and the technical skills they relied on to keep the boat operational and safe.

---

[58] Ragan, pg. 157

Crew changes like this can be dealt with, when there is sufficient time for the crew to learn to work together again and take each other's measure. Otherwise, there are problems – suspicion of 'the new guys', a lapse in performance as the new men are trained and become used to the routine. There is another point to keep in mind – Will Alexander wrote that by the time he left, he and Dixon were taking progressively greater chances with the *Hunley* - straining her endurance and speed to the limit, and often arriving back at Breach Inlet after the sun was up. Calculated risks with a crew that has worked, trained, and lived together is one thing, but extremely dangerous when one of the most experienced crew members is lost and replaced by three novices. Again, all of that can be overcome, but the crucial word here is time, time measured in weeks or perhaps months at best.

Hunley's crew had twelve days, and it would not be enough.

Alexander got his orders on the 5th and left that night. He wrote from Mobile as soon as he arrived to continue to offer whatever advice and assistance he could, and to encourage his friend to keep trying. It's unknown as to whether or not Dixon ever read them. It would have been another day or so before Carlson and Miller were transferred from Charleston over to Mount Pleasant, so it would have most likely been the 7th or perhaps as late as the 8th before Dixon took the *Hunley* out again. Within a few days though, it looked as if they might have a target that wasn't going to get away.

Back in September of 1862, a brand new Federal ship arrived to take up station off Charleston. She was the sloop-of-war USS *Housatonic*, built at the Boston Navy Yard in 1862 and commissioned that August. She came in at twelve hundred and forty tons, two hundred and seven feet long, low, wicked and as black as midnight in a coalmine. She was one of four ships in her class and was a state of the art warship, typical of a late Sail Age military vessel – fast at twelve knots, and an impressive punch of nine guns that encompassed everything from 24-lb howitzers to a massive 11" gun. The USN had been turning out ships like the *Housatonic* as quickly as they could, foreshadowing another war eight decades later when they were

The Long Patrol

doing the same with far more lethal ships, with equal surprise to other adversaries who were sure they couldn't do it.

"...Long, low, wicked, and as black as midnight in a coal mine..." USS Housatonic, the Hunley's target on the night of February 17th, 1864 as drawn by R.G. Skerret. (Photo courtesy of the United States Naval Historical Center, Washington, DC.)

Housatonic spent a quiet few months until the night of January 21st, 1863. That night her lookouts spotted a ship trying to get through the line, and they swung into action to stop her. Her gunners opened fire quickly and accurately, rapidly bracketing the target. It was a fast, well-drilled performance that was sullied only by the fact that it turned out to be another Federal vessel, taking up its place in the line. Needless to say, that did little for *Housatonic's* reputation in the Squadron and old Admiral DuPont could not have been complimentary to her skipper.

Fortunately, *Housatonic's* crew redeemed themselves just ten days later on the 31st. That night, *Chicora* and *Palmetto State* came out of Charleston to try and break the line. It was a gutsy performance by their CSN crews, outnumbered and outgunned, but there was a grim rationale behind it. Two days earlier, the runner *Princess Royal* had almost made it past the line, but was battered to a halt and captured within sight of Charleston's lights.

CSS H.L. Hunley

The loss had been far more painful than just that of the ship: she had been carrying *two* complete sets of British-built marine steam engines, heavy weapons, and ammunition, a load described by one writer as "the war's most important single cargo of contraband." Losses like that would kill the Confederacy faster than any defeats in the field would, and the line *had* to be cracked. Accordingly, the two ships went out that cold January night to possibly try to recapture the *Princess* – and teach the US Navy a lesson.

Early on the morning of January 31st, with a heavy fog covering the approaches to the harbor, the two warships slid quietly out, the lighter and more maneuverable *Chicora* in the lead. The first ship they encountered was the blockader USS *Mercedita*, which *Palmetto State* rammed and left sinking. Trying to reach *Mercedita*, USS *Keystone State* came out of the fog directly in the line of fire. *Chicora* and *Palmetto State* left her burning and dead in the water before taking on USS *Quaker City* and USS *Augusta* in turn.

The Federal ships, less well armed or armored than the two Confederates, were taking a beating before a black shape loomed out of the mist with its guns blazing – *Housatonic*, charging into the brawl at full speed, and this time her skipper was *very* sure of what he was shooting at. Within minutes, the sheer ferocity of *Housatonic's* assault had driven *Chicora* and *Palmetto State* back, and by that time the fog was starting to lift and the rest of the South Atlantic Squadron was heading for the fight as well.

The Confederates were courageous, but not suicidal – low on ammunition and facing rapidly mounting USN reinforcements, they put about and headed back for Charleston as quickly as they could. *Housatonic* didn't quit; she pursued them clear back into the harbor, turning back only when heavy gunfire from Confederate shore batteries started bracketing her.

From that day on, *Housatonic* had a reputation on both sides – a tough, aggressive ship to the Federals, and to the Confederates a vicious monster that had to be stopped. Her crew did nothing to diminish either reputation; she nailed the outbound runner *Neptune* with a valuable cargo of cotton and turpentine on April 19th, and did it again to the runner *Seesh* on May 15th. After that, *Housatonic* went in close to hammer Battery Wagner on July 10th, and was still there on the 18th to support the magnificent but ultimately futile charge of the 54th Massachusetts depicted in the film *Glory*. Once that was done, she moved onto the gun line that was firing into Charleston and its fortifications. General Beauregard and Commodore Ingraham might well have wanted to hang Admiral Dahlgren's flagship, the *Philadelphia*, on the Battery – but they would have happily settled for the *Housatonic*.

The ships in the line rotated positions from time to time, and at some point in early February, Dahlgren pulled *Housatonic* out of her slot and sent her to the northern end of the blockade line. There, her crew would be less likely to get into a slugging match with a blockader and she'd have time to get things tightened up – the little odds and ends that tended to slip a little bit during prolonged combat operations, as well as let the crew relax. As much as she probably should have been, *Housatonic* couldn't be sent back to Norfolk or Washington for drydocking and refit – she was needed on the line. So her wheel was put about one day and she moved about five or six nautical miles north of her old position, now sitting just over the horizon southeast from the quiet Isle of Palms.

And that was exactly where George Dixon saw her late one February afternoon as he scanned the horizon for prey.

It's not clear whether or not Dixon knew what ship this new target was, but its possible. *Housatonic* was one of the larger vessels on the line, and she was a purpose-built, steam-driven warship – she would have stood out from the converted merchantmen that made up the bulk of the blockaders like a nuclear powered aircraft carrier in the middle of a flotilla of tramp freighters. It's also likely that Dixon was seeking some kind of intelligence from Beauregard's or Ingraham's headquarters as to what was out there – that is the kind of detail Dixon would have wanted so to be as prepared as possible. But, as with so much of this story, we don't know for sure. We can be reasonably sure that Dixon saw her several days in a row, moving about slightly – just the tips of her masts over the horizon one day, the next her black bulwarks looming almost close enough to touch. And like any good submariner, Dixon understood that there were, after all, only two kinds of ships:

The *Hunley*, and targets.

The *Housatonic* had sat there, tantalizingly close, for some days prior to February 17th, 1864. It can only be assumed that on the afternoon of the 17th, following a couple hours attack practice in the Back Bay, Dixon went down to the beach below Battery Marshall and scouted potential victims – and there she was, closer than ever before, a slap in the face to everyone on that battered little island, a reminder that women and children in Charleston are starving under its gaze and dying under its guns.

Those would have been reasons enough, but there may have been one other to spur Dixon on. At this point, he has been in command of the *Hunley* for just about sixty days. He has taken her out on about thirty of those days, and has brought her back without a single victory to show for it. It has not been for lack of planning, effort, or courage on the part of George Dixon and his crew. They have been held back by the very real constraints of weather and the basic design limitations of the *Hunley* itself.

General Beauregard has been most understanding, as he knows and respects George Dixon and cannot possibly question his courage or dedication. But facts are facts: the *Hunley* has been on duty in Charleston since the beginning of August, and it is now the middle of February. Beauregard may not have said anything directly, but its hard to believe that word didn't filter down to Breach Inlet that the General was beginning to get impatient. And in an era when Honor was all, Dixon had to face the fact that his men had not delivered a single blow against the enemy. As the sun started to head down around 1800 hours that evening, Dixon had made up his mind. They were going out tonight, and they were going to kill a Federal ship.

Dixon was going to be violating one of his most sacrosanct rules in order to do so: it was a very clear, moonlit night. The *Hunley* would be taking a terrible, terrible risk out there - even though the target did not seem to have shown a calcium light (she was supposed to have one, but it's questionable as to whether or not she was using it), the moonlight alone would have been enough to expose them. But it appears to have meant nothing to George Dixon as he trotted back to the Breach Inlet base, waving his hat over his head to get the crew's attention.

One way or the other, tonight the *H.L. Hunley* was going to war.

Episcopal Bishop Richard H. Wilmer of Atlanta was asked during the Civil War to write a prayer for the Confederate States Navy. Bishop Wilmer sat down and gave the matter no little thought and then came up with this prayer, the original of which now resides in the Emory University Library:

"...O ETERNAL Lord God, who alone spreadest out the heavens, and rulest the raging of the sea; who hast compassed the waters with bounds, until day and night come to an end; be pleased to receive, ..."

We don't know if George Dixon said a prayer as he prepared the Hunley that night, but he would have been only human if he did, watching the men of his boat climb in.

Miller.

Becker.

Simkins.

Collins.

Ridgeway.

Carlson.

White.

Wicks.

Dixon would be last, looking at the two lonely sentries on the dock at Battery Marshall. Their departure was a regular thing now, and no one else would have been there. It wasn't that they didn't care, far from it. It's just that it was routine for the gunners at Marshall, now huddled inside their revetments and barracks trying to keep warm on a February night.

"...Into thy Almighty and most gracious protection, the

CSS H.L. Hunley

persons of thy servants, the officers and crews of our fleet, and of all other vessels now engaged in active service. Preserve them from the dangers of the sea and the violence of the enemy..."

Dixon lit the candle as soon as he got in, and as it flared into life he looked back into the interior, already beginning to bead up with condensation. Everyone was quiet, and nervous smiles greeted him in return. Of course they'd be nervous, Dixon would have thought. He had been nervous too going forward at Shiloh, and almost on cue his leg gave him a twinge.

"...Give them victory in their various conflicts, and safety and success in all their undertakings..."

Three miles out, the watch on *Housatonic* could see quite clearly lanterns and fires at the northern end of Battery Marshall. They wouldn't have been able to see any more than that, but it was all right. It was nothing to worry about.

"...That they may be a safeguard unto the Confederate States of America..."

Dixon popped back up through the narrow hatch and got himself oriented one last time. His eyes swung out to sea, and the lanterns were still there – seemingly close enough to touch, but Dixon knew they were at least an hour and a half's long hard labor away. At the end of those ninety minutes though, there would be one less ship on the blockade line, a dozen fewer guns to fire into Charleston.

"...And a security for such as pass on the seas upon their lawful occasions..."

Picket boats from both sides knifed through the moonlit water, sometimes close enough to see one another. Sometimes there would be surprisingly good-natured taunts exchanged, sometimes just hard, grim looks. None of them had any idea that the world would be a different place when they went to sleep tonight.

"...May they return in safety to enjoy the blessings of the land, with the fruits of their labour..."

Dixon leaned back down and gave a quiet command to

The Long Patrol

Wicks, who in turn started turning the heavy iron crank. With grunts of effort, the men started it spinning, fighting inertia and the water. As the boat began to move, Dixon looked back up at the sentries, two – well, young men wasn't the right word, they were closer to boys, really, neither one in a uniform or even looking remotely military except for the muskets they wore slung awkwardly over their shoulders. Dixon gave them a brief smile and nod, and they returned it – then snapped to attention and fired off the best salutes they could, remembering that this man was an officer in the Confederate States Army. Dixon stood tall, put his shoulders back and sent the salute back so sharply it fairly whistled.

"...And with a thankful remembrance of they mercies, to praise and glorify Thy Holy Name through Jesus Christ our Lord. Amen."

Dixon held the salute for a moment, then faced forward again to conn the boat out of Breach Inlet, carefully threading the *Hunley* through the narrow channel. The two sentries stood quietly and watched as the boat moved steadily away, a black shadow slipping noiselessly through the blue-black-silver of the gently billowing ocean. There was a faint glow coming up from beneath Dixon until he brought the boat to a stop, and then closed the hatch. They looked on as *Hunley* paused for just a moment, then began to creep slowly forward once more and steadily began to sink, easing carefully down until she disappeared and the water's surface was clear and quiet once more. The two guards were the last Confederates to see the *Hunley* and its crew.

And now, the mystery. The names do not ring down through history as much as they drift down, whispers that flit past our ears like half-heard cries for help before they and their crews vanish forever. *Marie Celeste*, found abandoned and adrift but without a crew. USS *Cyclops*, a massive collier that disappears in the Caribbean without a trace in March 1918. And of course, the *H.L. Hunley*, lost at the moment of her greatest triumph on a cold February night in 1864 within sight of her home base and missing until May of 1995.

CSS H.L. Hunley

The known facts are straightforward and beyond any reasonable dispute. At 2045 hours local time on February 17th, 1864, Confederate States Ship *H.L. Hunley* attacked and sank the Federal sloop-of-war USS *Housatonic*. With the exception of a blue signal light seen approximately forty-five minutes later, *Hunley* vanished, taking her nine-man crew with her. It's what we don't know that has kept the riddle alive for nearly one hundred and fifty years.

Why? What sent the *Hunley* down that night, seemingly within arms reach of her victim? Was it a last, parting shot from the dying *Housatonic*? Was it a tragic accident at the last moment? Was she killed by the explosion from her own weapon? Was she unknowingly run down by another Federal ship rushing to assist the *Housatonic*? The questions add up so fast that the head literally spins trying to keep them straight. Even the facts we do know are often contradictory of each other and they in turn raise more questions. What I hope to do from this point is to suggest a logical and reasoned timeline and sequence of events to the last patrol of the *Hunley*, and perhaps explain what happened and why. I shall do so by asking – and hopefully answering – a series of questions.

What time was it?

We know the *Hunley* left Breach Inlet after sunset – but what time was that? According to the US Naval Observatory, sunset that night was at 1807 hours. Most accounts have *Hunley* getting underway at approximately 1900 – at the very least, an hour's difference and possibly more. How do we know as accurately as possible what time it was?

The time is crucial here, and the Observatory has a role in the discussion. In 1864 the Observatory was to a great extent responsible for figuring out what time it was, a mission that they perform with exquisite precision to this day. Using measurements of star movements, they were able to determine exact times. In turn, US Navy ships set their chronometers to the master clock at the Observatory. This is important to keep in mind because the only reliable times we have from the night of February 17th are from the US Navy – the Observatory sunrise/sunset times and even more importantly the event times given by survivors from *Housatonic*. Although there is no indication as to whether or not *Housatonic* ever actually put into Washington to get her chronometers set, she would have certainly set them to ships that had. In addition, the USN had another advantage in figuring out and more importantly knowing what time it was: the bells system.

The bells system is familiar to anyone who ever served in the Navy and almost incomprehensible to anyone who hasn't, but I shall endeavor to explain it. Each day was divided up into six four-hour watches. This enabled the limited number of men on a ship to be able to maintain the ship, eat three square meals, and still get sufficient rest to stay sharp. The bells system was set up because most seamen couldn't afford a watch, watches had a tendency to fail at sea anyways, and many seamen were sufficiently illiterate that they could not have told time in any event. The bells – sounded on the hour and the half hour – told each sailor what time it was and where they needed to be. They may not have been able to read a duty roster or tell time, but they could count bells. The result was that even the most illiterate sailor knew what time it was to within half an hour, and those who could tell time but didn't have a watch could be even more precise. By way of example, if we have *Hunley* leaving at 1900, that would be six bells on the dogs' watch and the attack at approximately 2045 would have been between the first and second bells on the first watch.

On the other hand, the population of Charleston – civilian and military - had no access to accurate time measurement since April of 1861. There were no former USN vessels with chronometers in the Harbor to match their clocks up to and even if any CSN vessels had chronometers, after three years with no updates they would have been growing more and more inaccurate with every passing day. Civilians and CSA personnel ashore had it even worse. You could, of course, set your watch to noon by simply waiting for the sun to go directly overhead and the shadows to disappear or the bells of Charleston's church steeples to ring – but neither one of these is conducive to precise timekeeping. As old style mechanical watches wear on, they

CSS H.L. Hunley

become progressively less accurate, and the results are a pretty wide variance of times.

Having said all of that, I believe the approximate 1900 departure time is accurate and can be used as a baseline. This in turn is based on the most solidly determined event time of the entire evening: the detection of the *Hunley* at 2045. As explained above, the *Housatonic's* crew had a pretty solid grasp of what time it was. What we need to know now is *Hunley's* speed.

Although *Hunley* was capable of up to 4.5 to 5 knots, that was her optimal performance with a well rested crew and a boat in brand-new condition. We already know that *Hunley's* crew was far from rested or healthy, and we can be reasonably sure that the boat had not gotten any serious maintenance since being beached in November. What is much more telling, however, are the estimated speeds reported by the crew of the *Housatonic*. In the official final report on *Housatonic's* loss, *Hunley's* speed during her final approach is estimated by knowledgeable, professional naval officers to be no more than 3 or 4 knots – this with the crew literally cranking for their lives. Four knots during the few moments of the final approach is more than reasonable – but I do not believe it was a sustainable speed during the trip outbound from Breach Inlet. What seems much more likely is an average speed between 2 to 2.5 knots. Add to that the ebb tide – around one half to three quarters of a knot – and they were probably averaging 2.5 to 2.75 knots.

Housatonic was 2.7 nautical miles from Breach Inlet. If we work with a twenty-minute submerged running time with a five-minute surfaced and stopped air recharge time, departing Breach Inlet at approximately 1900 puts them at the *Housatonic* around 2030 – plenty of time for Dixon to line up his approach and strike at 2045.

What Were The Weather Conditions?

For reasons comprehensible only to the Federal Government, weather records for Charleston only go back to 1912, but what we have is more than enough to work with in terms of giving us an accurate idea of what the crew faced that night. We do know the average temperature in Charleston in February is about 35.5 degrees, with twenty days of the month below 32 degrees - that is, it is quite likely that it was below freezing or headed rapidly in that direction when the *Hunley* shoved off from the Breach Inlet dock. We know from the Galena letter that the sky was cloudless and clear, with some pockets of mist visible, and there was little or no breeze at all. The moon was in its first quarter[59] - not, of course, as bright as a half or full moon, but with a clear sky this would have made for good visibility at sea. Overall - with the exception of the potentially lethal cold - the weather was one of the few things going in Dixon's favor that night.

What was the condition of the crew?

As discussed earlier, it was not good. This might be a good time to review in more detail what physical and mental handicaps might have been crippling the crew of *H. L. Hunley*.

Their diet was, by our standards, abysmal. The crew was getting – usually once a day – a few small pieces of salt pork, some wild rice, locally picked greens, and hardtack. Very occasionally there may have been some eggs, oysters, or fish. But regardless of what occasional treats they may have gotten, in general their caloric intake and nutrition were abysmal compared to what we know are healthy levels. They were badly deficient in most vitamins and minerals, and the list of diseases and afflictions they were susceptible to – and probably suffering from - is staggering. The chart below shows approximately what they were getting from their rations, and what it was doing to them.

Category	% Of Modern RDA	Possible Effects of Deficiency/Excess
General Calories	27% (based on 2400 kcal)	Loss of muscle mass and strength, bone loss, cachexia

---

[59] http://eclipse.gsfc.nasa.gov/phase/phases1801.html

Vitamin A	143%	Headache, peeling of the skin, enlargement of the spleen and kidneys, bone thickening and joint pains
Thiamin	33%	Beriberi, with increased potential for heart failure, abnormal nerve and brain function
Riboflavin	26%	Fissures and scaling of the lips and corners of the mouth, dermatitis
Niacin	32%	Pellagra (dermatosis, inflammation of the tongue, abnormal intestinal and brain function)
Vitamin B6	34%	Anemias, nerve, and skin disorders
Folic Acid	63%	Decrease in the number of all types of blood cells, anemia
Vitamin C	71%	Scurvy (Inflammation of the gums, bleeding, loose teeth)
Vitamin E	37%	Rupture of red blood cells
Calcium	26%	Muscle spasm
Iron	60%	Anemia, difficulty in swallowing, spoon-shaped nails, intestinal abnormalities, impaired learning ability

Magnesium	40%	Abnormal nerve function
Phosphorous	64%	Irritability, weakness, blood cell disorders, abnormalities of intestines and kidneys
Potassium	39%	Paralysis, heart disturbances
Sodium	160%	Confusion, coma
Copper	182%	Copper deposits in the brain, liver damage

Keep in mind that by no means am I suggesting that the *Hunley's* crew suffered from all of these ailments, nor were they suffering from the worst-case effects. But these diseases and effects, in part or in total, would have in sum been terribly debilitating at best.

It should be noted that this was the daily diet for the vast majority of the Confederate defenders of Charleston, who would also have been suffering from the same problems. But those assigned ashore and to ships had advantages the *Hunley's* crew didn't.

First, CS officers, especially those aboard ship, often pooled their own funds to buy extra and more nutritious food for their men. Even an officer as frequently dismissive of his men as George Custer contributed to the greater good when the Government would not. Units could also partake of the time-honored practice of foraging – sometimes buying extra food from the locals, sometimes begging it, sometimes stealing it. Notice also the possible effects that their diet was having on them. There was no such thing as nutritional science in the 1860s, and in many instances all anyone knew was that fresh fruit prevented scurvy. Anything else was a mystery, and a potentially fatal one at that.

Hunley's crew had neither of these luxuries. Dixon and Alexander together probably did as much as they could on the

CSS H.L. Hunley

salaries of two junior Lieutenants, but it could not have been much and after Alexander transferred out it was even less. Foraging took time and effort – and as *Hunley's* crew was performing its duties nineteen hours out of twenty-four, there simply wasn't any time. Finally – and perhaps most importantly – although men ashore and afloat were under fire, they could become accustomed to that. However, they did not have the daunting physical and mental effort that the crew had to face every day – a two-hour march, nearly twelve hours of cranking the boat, and the nerve-racking experience of being locked in the stinking, claustrophobic hull, cranking several nautical miles out and then slipping into the blockade line, sometimes close enough to hear their enemies talking and singing.

They were also only getting about three to four hours of sleep daily, with all the attendant problems of sleep deprivation. The brain's most basic capabilities are greatly impaired, and problem-solving abilities are badly crippled. Thought processes become locked into an inflexible series of reflexive actions. Emotional mood begins to disintegrate, while stress and anxiety levels begin to reach stratospheric levels and create a brutal, steadily descending cycle of illness. It becomes difficult if not impossible to perform the most simple tasks, like performing basic motor functions or even getting one's eyes to focus. Most importantly, endurance and simple strength start to fall to levels where the ability to merely function in civilian life becomes problematical. Survival in combat or possible combat becomes unlikely.

Respiratory problems – aggravated by the cold, wet conditions inside the boat, the outside temperatures, and the close quarters the men shared both inside and outside the boat, would have become a chronic, unending battle for the men. This would have also weakened them further and progressively destroyed their lung capacity – they couldn't have properly used what air they could get inside the boat.

Medical care was difficult to come by under the best of circumstances. This was still the Heroic Age of Medicine - no antibiotics, no understanding of nutrition or stress, and that is *if* there was a doctor available. The Confederacy was as shorthanded with doctors as it was with everything else, and those physicians who were available were for the most part with CSA forces in the field. Again, Mark Ragan's encyclopedic research points us in the right direction here – there is no indication that they ever saw a doctor for any reason.

There is also the likelihood that even if there was a doctor immediately available – and there should have been one a mile or so down Sullivan's Island at Fort Moultrie – the men may not have availed themselves of him. These were not men used to seeking medical care in any event except for immediate and fairly serious injury or imminent death. If they were suffering from a series of comparatively low-grade problems, they would most likely have simply decided to tough it out until they became incapacitating. These were men who were utterly dedicated to their commander and their mission – under the moral standards of the day, seeking medical attention that might have kept them from going out could have been construed as cowardice, and the crew of *H.L. Hunley* was never going to allow anyone to make that charge.

By way of comparison, the crew of the *Housatonic* was living in fairly luxurious conditions. For decades, the US Navy had been dedicated to maintaining the health and safety of its crews, recognizing that if the crew cannot function, neither can the ship. Although their overall diet suffered from some of the same deficiencies as that of the *Hunley's* crew, there were some significant differences.

First, the men at Breach Inlet were only getting one solid meal a day, with an occasional handful of food once they were out to sea. On the other hand, US Navy crews received three meals a day heavy in fruits and vegetables that alleviated much of the damage.

Although there was certainly a fair amount of stress involved, on the blockade line it was far less concentrated and was in the open rather than the stifling confines of the boat. The USN's watch schedule insured at least eight hours rest for a sailor every day, and there was reasonably competent medical care available on just about every ship in the blockade line. On smaller ships that didn't have a surgeon assigned, it would take only a short time to get an ill or injured sailor to a ship that did have a doctor aboard.

One more advantage that USN crews could enjoy was the fact that they could leave. Men went on leave or furlough on a regular basis, and ships rotated in and out of the line all the time. (Admittedly, *Housatonic* had been there for more than a year, which seems unusual. It's likely that since she had proven herself to be a reliable ship, the USN commanders there kept extending her tour – proof once again that some things never change. During Operation Desert Storm, the carrier USS *Saratoga* (CV-60) ended up on the line for more than a year – a modern record.)

The crews of the *Hunley*, however, had nowhere else to go. To come off the line was to retreat, to abandon your very home. They were assigned to their boat literally until either victory or death.

In summary and in short, at best these men were badly undernourished, suffering from serious nutritional deficiencies, and slowly deteriorating from sleep and stress-related problems. At worst, they were on the thin edge of physical and mental collapse and never should have been in the *Hunley* that cold February night.

What was the condition of the Hunley?

The boat was operational in the most basic sense – that is, she was able to move under her own power, dive and surface on command, and steer where she was pointed. Having said that, it must be repeated that as nearly as can be determined, there is no indication that the *Hunley* ever got any kind of care other than the most basic pre-operational checks between the time she went back into the water in early December and February 17th. There are two specific points that should be addressed here: first, the boat's stuffing boxes – the points where the boat's control rods and propeller shaft exited the hull – and the *Hunley's* Sunday punch, the hundred and thirty-five pound Singer torpedo.

The stuffing boxes were small, literal boxes at four points

inside the hull: where the rudder control rod, the propeller shaft, and the two diving planes actually penetrated the hull. Only two of these – the diving plane boxes – were actually accessible with any relative ease, as they were inside the main hull, within arms' reach of the crew. On the other hand, the rudder control and shaft boxes were behind the aft ballast tank bulkhead, and underwater when the boat was in the water.

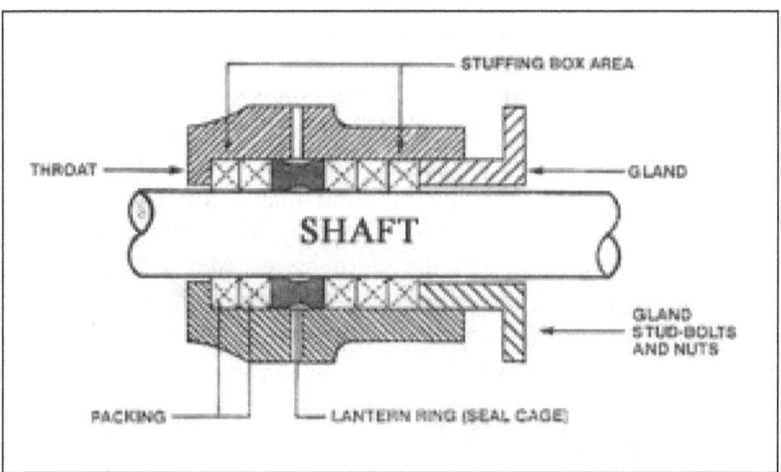

On *Hunley* these boxes were stuffed with heavy yarn – actually closer to a thin rope – as tightly as possible and when the boat was in the water, the stuffing would absorb any leakage. It worked surprisingly well, but it had to be carefully watched and re-stuffed on a regular basis. Otherwise, the yarn would eventually deteriorate, break up, and allow water to enter. Given the amount of time without any major maintenance, I strongly believe it possible and indeed likely that the stern stuffing boxes – rudder controls and propeller shaft – had not even gotten anything but the most cursory checks, if that, from the time *Hunley* went back into the water in January to February 17th.

Any leaks in the diving plane boxes, easily visible and accessible in the hull, would have been quickly spotted and re-stuffed. The stern boxes, behind the aft ballast tank bulkhead, would have been barely visible from inside the hull and probably inaccessible unless the boat was completely pumped out and brought ashore. Given the stresses the crew was under and the limited time they had each day, I think human nature steps in here – if the boat was not showing any obvious ill effects, they may very well have simply assumed there were no problems and pressed on accordingly. With that in mind, I believe there is a very strong chance that *Hunley* was actually leaking slightly through the stern boxes for some time prior to February 17th.

Progressive, non-fatal flooding is not unknown in naval history. One of the best examples is that of the battleship USS *Pennsylvania* (BB-38). Struck by a Japanese torpedo in mid-1945, she was never fully repaired due to a lack of facilities and the imminent end of the war. When she was sunk more than a year later during a nuclear test, she was still shipping water from the torpedo hit – and that was with zero maintenance being performed. It is with that in mind that I consider it highly probable and indeed likely that *Hunley* was taking on small amounts of water through the stern boxes, and this would be aggravated by the events of February 17th. But as mentioned above, if it wasn't causing a noticeable list or settling, it probably would have been ignored.

None of this is intended in any way to suggest that George Dixon and his men were lax or incompetent in taking care of the *Hunley*. They had one real sailor aboard in the person of Bosun's Mate Wicks, and he would have certainly done his best to maintain the boat in some semblance of good order. Dixon and the rest of the crew certainly would have been on the lookout for anything that was a potential problem, and would have fixed it had they found it. But they were running the world's first combat submarine – there had been no long training period, no carefully written manual or checklists. They had simply gotten in the boat and taken off. They could only do their best to keep an eye open for any difficulties, some of which might not have become apparent until it was far too late.

The other maintenance problem would have been with the Singer Torpedo. To review – it was a closed cylinder made of sheet copper or tin, large enough to hold one hundred and thirty-five pounds of black powder. It had a spring-loaded firing device that was triggered from the forward hatch, and was equipped with an iron or steel barb that was intended to keep it firmly lodged in the side of any unfortunate target. The actual

The Long Patrol

blueprints for the *Hunley* torpedo were found after the war and then copied for Federal records, and they seem to establish pretty solidly that the torpedo was carried on the end of the spar and not slung under it.[60]

There was certainly no question about Mister Singer's workmanship – Singer torpedoes had been and would continue to be responsible for the destruction of Union vessels until the end of the war. But the manner in which the *Hunley's* torpedo was carried, used, stored and maintained suggests the strong possibility of a misfire or an actual failure to fire at the moment of truth.

First, the torpedo was removed from the *Hunley's* spar at the end of every patrol and brought ashore. The question is, what was done with it? Was it simply left on the dock at Breach Inlet, or was it stored in a manner more keeping with its destructive potential? It makes sense that it was probably carried the few hundred feet to a magazine or revetment at Battery Marshall – the cannon stationed there would have required a magazine of some sort to maintain ammunition, and they may have even had an Ordnance specialist assigned to the Battery as well. William Alexander's later writings confirm that indeed, this is what they did. Well and good – the torpedo is in safe and secure hands until it is called for after the day's training runs in Back Bay. But is it in good condition?

Once again, we are faced with a dilemma – no indication of any kind that maintenance or upkeep is being performed on the torpedo. Remember:

- The torpedo is submerged in the water every day from sunset until the time the boat arrives back at Breach Inlet. This has been the routine from at least mid to late January.

- We are reasonably sure that the boat has grounded at least once and possibly more – which means the torpedo,

---

[60] Ragan, pg. 176. Scharf also has a drawing of a similar torpedo on pg. 754, but that one seems to be made of wood. However, the spar mounting attachment is identical to that of the Singer torpedo Ragan shows.

mounted on its twenty-foot spar, has been shoved into the sand off Sullivan's Island – with the risk of attendant damage.

- Carrying the torpedo by the barb – the easiest way to pick up the torpedo (the other end could have been easily carried simply by sliding a stick or rod into the spar attachment point) - would have put stress on the relatively light metal construction of the torpedo's forward end, along with the fragile spring-loaded trigger assembly mounted just behind it. In addition, a barb carry may have loosened the safety pin that kept the spring in place. The diagrams do not appear to show any kind of secondary safety device for use after the torpedo was removed from the *Hunley*, so the firing line and its reel – mounted to the hull slightly forward of the forward hatch – had to have gone with it, which may well have made carrying the torpedo a three-man operation.

- There is no indication that there was more than one torpedo. It is indeed possible that there was, but the evidence seems to point to only one – which means there was no replacement in the event of damage or maintenance of the torpedo. This means that the torpedo was there until either death or victory, just like the crew – and was taking its own unique beating every day.

Given the above points, there is a serious risk that the riveted copper structure has failed at one or more points through corrosion or stress. Seawater has leaked in, soaking and contaminating the black powder explosive filler. That in turn means that there is a strong risk of corrosion or damage to the most fragile and delicate part of the entire torpedo: the spring-loaded trigger assembly. Whether the spring was made of copper, thin iron, or steel, the constant exposure to seawater would have threatened it with corrosion, and constantly being under tension would have risked weakening it through metal fatigue. There is also the possibility mentioned earlier that on impact the torpedo could suffer serious or catastrophic structural failure as it is slammed against the side of its target. I believe all these things not only could happen but also *did* happen, and help explain the loss of the *Hunley*.

By the same token, *Housatonic* was in excellent shape. Although she had been on the blockade line for over a year, it had not been terribly hard duty. For the most part, she had been anchored offshore with a few forays onto the gun line or chasing runners. Her engines, a powerful steam plant built by Globe Iron Works in Boston, were in superb condition. The performance of steam plants had advanced impressively over the preceding few years, and they were no longer the balky, marginally capable units they had been just a decade before. The official reports and statements of *Housatonic's* crew state unequivocally that in the few moments between the order being given and the actual torpedo impact, the Chief Engineer was able to get the engines moving in reverse – and seemed to have no concerns about his ability to do so. This strongly indicates that the Chief Engineer was utterly confident of the condition and capabilities of the engines.

Keep also in mind that the vast majority of a warship's crew is there solely to maintain the ship, and aboard *Housatonic* they had a great deal of work to keep them busy. *Housatonic's* skipper, Captain Charles Pickering, was a tough, no-nonsense old salt that liked poking around his command and anything that needed to be done would have been taken care of immediately if not sooner.

The crew could also be considered in excellent condition as well, and not merely in a physical sense. Pickering had them trained within an inch of their lives, and *Housatonic's* performance and reputation was proof of that. In addition, the final report of *Housatonic's* loss made it crystal clear that:

"...The watch on deck, ship, and ship's battery were in all respects prepared for a sudden offensive or defensive movement; that lookouts were properly stationed and vigilance observed, and that officers and crew promptly assembled at their quarters."[61]

---

[61] *Official Records*, Series II, Vol. I, pgs 332-333

CSS H.L. Hunley

In summary – United States Ship *Housatonic* was in fighting trim, with a trained, capable crew and a skilled, experienced Captain, ready for anything...or at least anything that they had been *trained* to expect.

What happened during the actual attack?

For the sake of clarity, I will consider the attack itself as being the period from approximately 2030 to 2055 hours on February 17th, 1864. This way, even allowing for minor discrepancies in event times between witnesses on the *Housatonic*, we still get a fairly precise block of time that covers everything that took place and allows us to use it as a foundation and framework for post-attack events as well.

On USS *Housatonic*, it is 2030 hours – one bell on the first watch, 8:30 PM (more or less) ashore. Sunset was two hours and twenty-three minutes ago, and twilight ended twenty-five minutes later. Captain William Pickering is in his cabin aft, finishing his daily paperwork. His executive officer, Lieutenant Francis Higginson, is on deck as is Officer of the Deck John Crosby. Quartermaster John Williams is on duty as well, and another officer whose name is unknown was on watch with them. There are six more enlisted lookouts distributed about the deck and rigging, all of them watching for anything that could be a runner or a torpedo boat. The officers have spyglasses or binoculars to assist them, and we may assume the enlisted lookouts did too.

The rest of the crew is easing into the nightly routine. The galley and wardrooms have finished cleaning up after the evening meal, while most of the crew is enjoying some brief personal time before lights out. The rest are at their stations – some able to relax reasonably warm belowdecks, the rest trying to keep warm topside. The most important of those are the gunners, who have the most lethal and up-to-date heavy weaponry in the Navy at their disposal. Every weapon is cocked and ready, with ammunition and powder positioned beside it to keep them going until the powder monkeys start bringing up the rest from the magazines. The two biggest weapons – the 11" Dahlgren and the 100-pounder Parrott rifle are trained out to port

The Long Patrol

and starboard respectively. The Parrott is the only weapon facing the *Hunley* whose bearing can be changed. It is more than capable of sending the *Hunley* to the bottom with even a near miss.

Housatonic is at anchor, swinging gently with her bows bearing east-southeast. She is not entirely at rest – in the engine room, Assistant Engineer Mayer is keeping a watchful eye on the machinery. He is maintaining twenty-five pounds of pressure in the boilers - more than sufficient to get *Housatonic* underway, albeit slowly. The twenty-five pounds of pressure is part of Captain Pickering's standing orders while they are at anchor. He has no intention of allowing his ship to end up like the *New Ironsides*, rattled so badly by a sneak attack that she had to come off the line. In addition, Captain Pickering has instructed that in the event of an emergency, the deck crew is to slip the anchor chain and the duty engineer will immediately give the engines full *astern*.

It seems that some time before, Pickering had needed to get underway quickly, so he gave the order to slip the anchor and go full speed ahead – and the ship, moving forward, fouled the slip rope around the propeller. Since such an event rendered a warship momentarily useless as well as being most embarrassing, Captain Pickering has insured that it will never happen again. He may have also insured that his ship will be blown apart in fifteen minutes.

As the bells sound across the cold blue swells, *H.L. Hunley* is on the surface with Lieutenant Dixon peering through the forward hatch deadlights at his prey. The trip out has been smooth and steady, with no problems to delay them. Dixon has been able to keep the *Hunley* locked on course and has brought her out in very nearly a straight line from Breach Inlet, no mean feat. She is roughly six hundred feet away from *Housatonic*, whose lights are clearly visible and whose form sits on the calm waters like some malevolent sea creature threatening Charleston. *Hunley's* torpedo spar, winched up to about a twenty-degree angle, is aimed directly at *Housatonic's* starboard side. Dixon appears to have planned a slow, careful approach, intending to get as close as he possibly can before sprinting in for the kill.

CSS H.L. Hunley

The trick will be to *get* that close - based on witnesses' testimony, average visibility is around two to three miles.

With the previous torpedo boat scares, the United States Navy is on the alert and ready. Anything moving steadily will be seen and fired on, and while the *Housatonic* may kill at more than a mile the *Hunley* has a tactical reach of just about twenty feet. With silent motions of his hand, Dixon gives the crew the order to start cranking...very, very slowly. The shaft will rumble in its bearings as the men give whatever effort they have left, and *Hunley* begins to glide forward. They turn for just a minute or two, barely moving, before Dixon orders all stop again and they drift silently while he makes a few gentle course corrections.

Every inch closer they get, no matter how stealthy they are, the possibility of detection goes up exponentially. It is possible that Dixon is very carefully taking *Hunley* down, then bringing her up again, and this is indeed a smart, logical approach. But it is also insuring that the air is not being recharged sufficiently as the crew, nervous and exerting themselves, is depleting what air is available. If they are as ill as it is possible they may be, what air they are getting is not being fully absorbed by their lungs. No matter what, the CO^2 levels inside the hull are already soaring. Every breath is increasing them, preventing the transport of oxygen through their bloodstreams. Modern standards call a 4 percent CO^2 level 'dangerously unhealthy'. The crew of the *Hunley* probably exceeded that level several times a day, and were paying the price.

From 2030 through 2044, everything on the *Housatonic* is absolutely, perfectly, and completely normal. There is no reason to assume that anything out of the ordinary is going to happen. Lieutenant Higginson is pacing slowly from one side of the deck to the other, his eyes tracking and checking each of the other lookouts. The men on deck are huddling in their jackets, or talking quietly as they wait for nothing more than the end of their watch.

Somewhere around 2044, Dixon has *Hunley* on the surface roughly one hundred yards from *Housatonic*. It has been a

The Long Patrol

flawless approach, skillful and worthy of the men who will do the same thing with bigger, faster boats eight decades later. Dixon has gotten his boat and his crew not only into the blockade line, but within three hundred feet of one of its largest and most powerful vessels. And it is all about to go terribly wrong.

At 2045, Higginson – on the port wing of the bridge – pivots on his heel and faces starboard again. Everything looks fine, but Quartermaster Williams is standing on the deck, arms akimbo and his head slightly cocked, looking at something off to starboard. John Crosby has seen it too, and Higginson strides quickly to the starboard side. As he does so, both Crosby and Williams bring their glasses up to take a look. It's not clear who spotted – well, *something* first, Crosby or Williams. Crosby was the only one who ever made a claim, and he states that he saw something that looked like a porpoise, while Williams said he saw nothing. The Unknown Officer tells us in a letter to what was apparently his hometown newspaper that Williams spotted it first, and dismissed it as a school of fish.[62]

That misidentification bought *Hunley* a few more seconds, but it's not clear whether or not Dixon knew it. He was still slowly easing the boat towards its target, though at this point, it seems she was still drifting on the surface. However, if at least two separate observers saw her and mistook her for a fish or school thereof, she must have been ballasted down as far as Dixon could possibly get her, and may have been just high enough for Dixon to get the forward deadlight clear or perhaps just high enough to get the hatch partially open.

Whoever saw it first, Higginson was almost certainly the one who realized that something wasn't right. The Lieutenant saw something that looked like a 'plank sharp on both ends' (*Hunley's* forward cutwater and the water streaming back behind the aft hatch) heading directly for them. Planks most definitely did not move with that kind of speed or deliberation, and they don't have the faint illumination that Higginson also thought he saw (though it is possible that what he was seeing was natural

---

[62] See Appendix 4.

phosphorescence created by *Hunley's* movement through the water, and not the candle inside the hull.)

The Lieutenant didn't need to reflect on the matter any further. Higginson gave the order to 'beat to quarters' – a drum roll that's roughly equivalent to today's General Quarters or Battle Stations – and orders the chain slipped and the engines given full speed astern. *Housatonic's* crew, well trained and thoroughly drilled, sprang into action. Captain Pickering heard the commotion and immediately headed topside, grabbing his personal weapon – a double-barreled shotgun – as he left his quarters.

Immediately climbing up to the bridge, Pickering asks Crosby what in blazes is going on, and Crosby points to starboard. Pickering spots it immediately and later states that it looks like a plank and is moving at between three and four knots. That's enough for Captain Pickering – he levels his shotgun and opens fire on the *Hunley*, joining the other sentinels on deck who are now emptying their rifles and pistols as well.

Whether or not Dixon is tracking *Housatonic* through the deadlight or an open hatch, he had to have seen the sudden commotion on his target's deck - with everyone pointing and looking at *him* – and he knew he'd been spotted. There was only one option at this point, and Dixon took it: he ordered full speed ahead. The crew reached down into themselves to find one more burst of strength and determination, then started turning the crank with everything they had.

It is apparently only after he starts moving to make the final run that the small arms fire begins. It is suggested that at this point (and the film *The Hunley* takes this tack), *Hunley* takes a small-arms hit, most likely from a rifle, that shatters the forward deadlight or the hatch coaming and either injures or incapacitates Lieutenant Dixon. I do not believe that to be the case, for reasons we'll elaborate on later – my feeling is that *Hunley* goes into the attack completely undamaged except for buckshot and stray rifle rounds caroming harmlessly off her black iron hull. At this point it is roughly 2046, and it must seem that things are almost happening in slow motion. *Housatonic's* crew is moving

The Long Patrol

as fast as they possibly can to their battle stations, while the gunners on the 11" Dahlgren are getting the weapon unlimbered. Captain Pickering roars out orders to slip the chain and back astern and a brawny Bosun's mate swings a massive sledgehammer and knocks a wooden pin out from the chain, letting it rattle loose and freeing *Housatonic* to move.

In the meantime, Assistant Engineer Meyer has slammed the engine controls to full speed astern, and the huge steam engine thunders into life. The massive bronze propeller beneath *Housatonic's* fantail begins to spin, then bite, and the ship begins to inch backwards. This may be the action that actually dooms *Housatonic*, for as she begins to ease astern, her main ammunition and powder magazine is slowly being pulled directly in front of the *Hunley*.

The *Hunley's* crew is turning the propeller crank as fast as they can, and Dixon is holding the tiller with an iron grip. It has taken a few seconds for her to get moving, but she is now lancing forward like an avenging angel. Dixon can clearly hear small arms fire pinging off the hull, but he's seen far worse. His concern now is that massive Dahlgren rifle that now seems to be pointing directly at him, and he would have been able to see its crew trying to train the rifled cannon on his boat. As it turns out, he is too close and moving too fast for the gun crews to even get a shot at them. On the bridge of the *Housatonic*, Lieutenant Higginson watches Captain Pickering let another volley go at the *Hunley* as it begins to disappear from view beneath his ship's bulwarks. By most accounts, just about two minutes have passed since the *Hunley* has been spotted.

The delicately built chronometer aboard USS *Housatonic* gives a few more ticks and moves to 2047. If it remains within *Housatonic's* grave not quite three nautical miles off Sullivan's Island, it probably still shows that time today.

The steel barb at the head of *Hunley's* torpedo punches through the copper plating that covers *Housatonic's* sleek hull beneath the waterline, then splinters and tears its way through the oak planks that lie beneath it, but it is being twisted and pulled as the cruiser picks up speed and slides backwards, flexing the spar to the left – but also bending the barb itself, beginning the structural deformation of the torpedo. The barb has penetrated *Housatonic's* hull just forward of the ship's aft, or mizzen, mast – and directly in line with the ship's magazine, which has been moving steadily across *Hunley's* course as the ship moves in reverse. Captain Pickering will not have the embarrassment of fouling a line again, but his standing orders have inadvertently exposed his ship's most vulnerable spot.

As the boat continues on, sheer momentum pushes the spar into the rear of the torpedo. It is now being both compressed and pulled to the left, and the thin sheet metal structure cannot withstand the stresses. The torpedo probably did not catastrophically fail, that is, come completely apart and spill any of its contents – but what must be happening is the body of the torpedo compressing like an accordion and simultaneously twisting. Also, the torpedo's trigger, installed immediately behind the barb and mounted down its centerline, is taking the same beating – and being as delicate as the workings of a watch, it will not withstand the forces anywhere near as well. As the firing pin itself is pushed into the base of the barb, it flexes as well – and the delicate spring that holds it in place is either damaged or displaced; now becoming more of a hair trigger than a reliable mechanism.

While all of this has happened, Dixon has been slammed against the forward bulkhead, and the rest of the crew very nearly jolted out of their seats – but they recover quickly and react immediately to Dixon's bellowed orders for full speed astern.

As *Hunley* goes into reverse, the spar twists – or more likely breaks - out of the rear of the torpedo's body (causing the damage that can be seen at the end of the spar today) and the sub starts racing backwards. The gun crews are trying to train something, *anything* on the boat as it maneuvers away, but they aren't succeeding – the *Hunley* is too close and too small. Anyone with a rifle or pistol is banging away as best they can as the boat now clears the *Housatonic's* side.

Ensign Charles Craven, probably the last person aboard to

realize that something is going on, has now gotten up on deck and run to join the nearest gun crew – that of one of the four 32-pounders mounted on the starboard side. Their ability to train is limited, but Craven jumps in and they mange to manhandle the gun around so that it is trained on the *Hunley*, now about twenty feet away, still moving in reverse. Craven yanks on the lanyard, and *nothing happens* – the primer has misfired. Craven's actions in replacing the primer are probably the fastest in Navy history, and just before he pulls the lanyard again, he looks up.

Hunley is now about fifty feet back – Craven and his gun crew are probably the closest USN personnel to the escaping submarine, and it is very likely that they have the 32-pounder trained well enough to either demolish the *Hunley* or cripple it with a near miss. This and several other reports from *Housatonic* survivors are that last confirmed sightings of the *Hunley*. Craven will never know what his gun would have accomplished.

It is likely that Ensign Craven and the crew of the 32-pounder are either directly above or within a few feet in either direction of the torpedo when it detonates. We know that the *Hunley* was supposed to be much further away from the torpedo when it lets go, perhaps as far as a hundred and fifty feet, and at best *Hunley* is only a third of that. Research released in January 2013 suggests she may have been as little as twenty feet from the blast. We know that during the Mobile trials, the towed warhead with ninety-five pounds of powder aboard shook the *Hunley* hard, and that was a hundred and fifty feet away with the hull of the target between the boat and the explosion.

During whatever practice attacks they had run, Dixon was consistently simulating firing the torpedo at the same distances, and it's unlikely that the combat veteran would have intentionally fired the torpedo too early, especially when the hull of the *Housatonic* was still literally filling his vision. That leaves as the most likely possibility that the torpedo's firing mechanism has malfunctioned in some way.

This in turn breaks down to two main possibilities. First, the firing line has somehow snagged on something and yanked the safety pin out of the firing pin, allowing it to fire. This could have happened during the initial impact, when the rigging of the spar assembly may have broken and fouled the firing line. This is supported by the 2013 finding that the Hunley was only 20 feet from the explosion – the length of her spar.

Another scenario involves the torpedo spar itself. It was found more or less whole and intact, but it is ragged at the forward end. This suggests that the spar may have broken just aft of the torpedo, rather than pulled smoothly from the base of the torpedo. In doing so, the spar itself may have fouled the firing line. In either event, the firing line is tangled around something and is pulled tightly enough to activate the firing mechanism while *Hunley* is still far too close to the torpedo.

The other possibility is that the safety pin has been damaged – it may have fallen out entirely, or the distorted trigger assembly is keeping it in place. As the *Hunley* backs away, it will not take much of a jolt at all – a stray wave, or even the mounting vibration of *Housatonic's* own engines – to shake the spring loose, allowing the firing mechanism to activate.

- Ignition.
- Combustion.
- Explosion.

There is a bright green-blue flare through the water as the black powder filler detonates, but if the torpedo is as damaged as it should be not all the powder will go high order. It will, however, be enough to slice through the copper sheathing and splinter the heavy oak planks of *Housatonic*'s hull into matchsticks for a short distance around the penetration point. In 2013, fragments of the torpedo's casing were also found on the Hunley's hull, again suggesting she was very close to the blast.

What needs to be stressed at this point is that this single explosion should not, in and of itself, have sent *Housatonic* under. A Sail Age warship was a remarkably tough thing to kill – they were built from seasoned oak or pine, which was incredibly resistant to even close in, heavy gunfire – remember how USS *Constitution*, 'Old Ironsides' herself, got her name as a broadside from HMS *Guerriere* literally bounced off her flanks fifty years before – and designed to survive hours of similar pounding. Even the development of rifled cannon with explosive shells required massive damage to be inflicted on warships before they sank or were even knocked out of the line – again, remember the hits the Federal task force off Charleston took.

An excellent example was that of USS *Mississippi*, lost not quite a year before during the fighting to take down Vicksburg.[63] *Mississippi* was one of the first modern warships built for the US Navy – a sidewheel steamer of roughly the same dimensions and size as *Housatonic*, though twenty years older. She had sailed to Japan with Commodore Matthew Perry in 1851, and when the Civil War broke out she was in the thick of it from the beginning. But on March 14th, 1863, pride, courage, and history weren't enough to keep her from Gehenna. Second in line behind USS *Richmond*, *Mississippi* was pounded by the Confederate batteries at Port Hudson, Louisiana. Losing power and control, she ran hard aground, and despite the best efforts of her skipper and her Executive Officer – a young Lieutenant named George Dewey – they couldn't get her clear. The Rebels shelled her mercilessly, and her crew abandoned ship under a hail of heavy-caliber fire. Dewey calmly supervised a scuttling party that insured the ship was aflame when they left.

But as the ship flooded stern-first, the bow actually floated off the sandbar that had trapped her up to that moment. Now, consider this for a moment: up to this point, *Mississippi* has been taking repeated heavy-caliber hits from just after 2300 hours on March 13th until around 0300 on the 14th. She has been repeatedly holed beneath the waterline, taking on water, and is aflame from stem to stern – but she is *still afloat*. She would remain that way for two and a half more hours until the flames finally penetrate to her innards – and her main magazine. Forty thousand pounds of explosives cooked off, and even then it took some time for her to settle. When Clive Cussler located her wreckage in 1999, she was beneath eighty feet of swamp and sludge, but indications were that she was surprisingly intact.

---

[63] Cussler, *The Sea Hunters II*, pg. 116.

CSS H.L. Hunley

Mississippi was a typically built, typically designed, and typically maintained mid-19th century warship, which endured a hellish pounding that far exceeded anything her designers had foreseen, not actually sinking until after a massive magazine explosion finished her off. I believe that this is solid evidence that the best *sure* way to sink a Sail Age warship was to inflict a penetrating hit into the ship's magazine – exactly where the *Hunley's* torpedo had penetrated.

The detonation's incandescent bloom knifes through the disintegrating planks and into the magazine spaces, which are full of cloth bags of gunpowder, shells and cannonballs, and fuzes for those shells. Given 1864 explosive technology, using powder-based explosives and with an almost complete lack of any physical protection, any kind of unguarded flame or spark is a potentially ship-killing danger – a rogue explosion is a death sentence, and that is exactly what is read unto the *Housatonic* at thirteen minutes to nine on a cold February night. The powder bags – silk containers holding the powder charges necessary for the big guns topside – are most easily ignited, and they start the nightmare rolling.

The explosion that follows is the definition of the word 'catastrophic'. Explosions will follow the path of least resistance, and that is exactly what happens here, first bulging, then fracturing, then shattering the lacerated oak hull into a hole that would have been big enough to drive a truck through. The heavy brass propeller shaft, rapidly spinning up to speed as steam pressure mounted, is severed and the forward half rears upwards, through the overhead and into the decks above before the bearings seize.

The hull is now rupturing up, down and under the starboard side just forward of the cabin. Within milliseconds *Housatonic* has been transformed from a sleek, lethal greyhound of the sea into two mostly separated hulks, now both rapidly filling with the cold Atlantic Ocean. The explosion has blown Ensign Craven almost fifty feet *forward*, and when he is able to regain his senses he is close to the mainmast - in water up to his ankles. Captain Pickering is literally bounced across the deck, badly bruised and partially unconscious but alive. Lieutenant Joseph

The Long Patrol

Congdon, one of the other officers topside, has been firing on the *Hunley* with his revolver but the explosion knocks him to the deck.

When he stands up he can see nothing in his immediate vicinity and believes that Captain Pickering and Lieutenant Higginson are dead. He is fairly sure that he is the only surviving officer on deck right now, and it does not take years of experience at sea to know that *Housatonic* is gone, wounded beyond help. Congdon is mistaken about the surviving officers but it is easy to understand and he should not be chastised for it. On the other hand, he is absolutely correct in his assessment of *Housatonic's* condition: his ship is lost, she cannot be saved, and the only real option is to get the men off. Congdon gives the order – lower the boats, get the men off. The ship is now heeling to *port*, making any actions more difficult to accomplish by the second.

His actions are almost too late. The *Housatonic,* gutted and dying, is settling so quickly that the most of the crew will not even be able to get to the boats and find themselves still standing on deck as water begins to swirl around their legs. Congdon orders them to start climbing the rigging, and they do it with alacrity as the holystoned deck begins to disappear under the water. They are for the most part still there when the *Housatonic's* keel thumps down against the sandy bottom and she finishes filling with water, becoming one with the ocean she has so confidently patrolled. By all accounts, it has taken just about seven minutes from detonation to the *Housatonic* striking bottom.

For all the violence and speed of the nightmare that has just possessed the sleek cruiser, there have only been five deaths – Ensign E.D. Hazeltine, Clerk Charles Muzzey, Landsman Theodore Parker, Fireman Second Class John Walsh – and Quartermaster John Williams, who one can only hope did not have long to regret his error in identifying that thing a hundred yards away from his ship.

The crew knows they will not wait long, and those who could not get into the boats are now perched in the rigging – cold

CSS H.L. Hunley

and shivering, but alive. The boats – one carrying the injured Captain Pickering – are now headed for the nearest vessel, the sloop-of-war USS *Canandaigua*.

In the meantime, all Hell has broken loose aboard the *Hunley*.

Had it just been the torpedo itself, things would have been bad enough but they might not have been fatal. But it had all gone wrong in just the space of a minute or so – being spotted just a few seconds away from making the final run, *Housatonic* moving aft instead of forward or just sitting there, and a damaged torpedo having just enough punch to get through the hull timbers and ignite the explosives on the other side. And although the crew is cranking as hard as they possibly can, for all practical purposes the *Hunley* may as well be motionless, for the shockwave that its coming to take her life is moving at several times the speed of sound through the expanding hole in the cruiser's side. Semi-circular and expanding, it races across the short distance still picking up speed and strikes the *Hunley* like a giant's fist.

The science of hydraulics was in its infancy, and no one yet could even conceive of the idea of water moving in such a manner as to be almost solid in its effects. This however is exactly what happened to the *Hunley*. The shockwave wrapped itself around the little boat, squeezed, and shook it like a rag doll, physically propelling it further away from the dying *Housatonic*. The ovoid shape of *Hunley's* main hull probably saved the crew from being smashed and drowned out of existence right then and there, but the more squared-off bow and stern could not survive the sheer brute physical force that was being applied against them. Their solidly riveted hull joints would have probably even held up well – briefly - under direct fire, but the ironworkers at Park and Lyons couldn't even comprehend anything like this when they assembled the *Hunley*. The seams, riveted over stiffening formers, gave, ruptured, and failed.

For decades, people considered that when the doomed RMS *Titanic* brushed up against an iceberg fifty-eight years later, a spur of blue sea ice had penetrated her hull and opened her up for two hundred feet of her trim, dignified flank. But after she was located in the mid-1980s, it was discovered that there was no single hole, only a series of sprung seams along that two hundred foot length whose total open space amounted to no more than that of an average doorway. It is more than likely that this was what happened to the *Hunley* – no great fountains of water bursting in to quickly swamp the boat and send her down with her victim but a death of a thousand cuts, water seeping in a drop here, a trickle there. Had the main hull been shipping water, the crew would have seen it at once and plugged it or abandoned ship if they couldn't do that. But in the ballast tanks, difficult to see under the best of circumstances, leaks might have been all but invisible.

That was forward. Aft it may have been worse. As the shockwave passed over the *Hunley*, it would have not only compressed and sprung the forward seams but it would have snapped the sub like a whip from stem to stern. Shock damage like that also has a precedent, especially in the Second World War and on US Navy ships struck by mines in the Persian Gulf. Ships hit by bombs or torpedoes – or even near misses – suffered serious internal damage some distance from the impact site that could take them out of the war for months. This is where the stuffing boxes may have also contributed to the mounting disaster – with sodden, deteriorating yarn as the rigid control rods and propeller shaft shook inside the boxes and opened up more small leaks. If anything then, the damage astern of the aft ballast tank bulkhead may have been worse than that forward.

The crew may have been battered just as badly. Dixon, half standing, half crouching in the forward hatch, would have been shaken worse than anyone else, but held on to regain control of the boat. The crew at their stations on the crank would have been tossed about strongly enough to make them come to a stop. If they were fifty feet or more from the blast, they should have made it through without serious injury. If they were twenty feet out as the 2013 results suggest, its an entirely different matter. That close, shock from the blast would have caused internal soft tissue injuries to the crew and resulting in them bleeding from the nose, ears, eyes and less-mentionable orifices. In addition, they would have all been suffering concussion and possible internal bleeding. All of which adds up to them going into severe shock over the a period of a few minutes to an hour or so.

It's also a good bet that the candle would have gone out, a detail that may have significance later. What is certain, however, is that Lieutenant George Dixon, Confederate States Army, was able to watch the *Housatonic* head for the bottom and let his crew know that they had become the first submarine in history to sink an enemy vessel.

Now they had to figure out a way to get home.

We know that at this point, the nearest ship – USS *Canandaigua* – is a couple of miles away, and one of *Housatonic's* boats is making for it at best possible speed. Dixon, on the other hand, does not know this. No radar, visibility limited by the *Hunley's* position low in the water – he cannot take the chance that someone or something is on its way to get revenge.

It appears that he did something quite logical and reasonable: he gave the order to take her down and wait. It makes sense – if you don't know or can't be sure what the situation is topside, go below and wait it out. But Dixon doesn't know that *Hunley* has taken serious and probably fatal shock damage. Going deep is tactically the right thing to do, but from an engineering point of view it is an awful mistake. She is already taking on water and going under will only make it worse – but it is the only realistic option Dixon believes he has. Opening the ballast valves, he takes *Hunley* down – not too deep, perhaps just a few feet.

It is enough. *H.L. Hunley* and her crew have just about an hour, perhaps seventy-five minutes to live. The hatches slip beneath the waves, and Dixon gives the order for ahead slow. Did he have the candle lit? Probably, but I suggest that he did not do so until after they'd submerged – again, remember that Dixon was a combat veteran. He wasn't going to do anything to attract attention to himself, and he may very well have realized that the candle helped the *Housatonic's* crew spot him. In the meantime, the *Hunley* slowly crept away, trying to circle back to a north-northwest heading. The time is just about 2055.

The Long Patrol

What happened after the attack?

At about 2120 – 9:20 PM, the lookouts aboard *Canandaigua* report boats approaching from the south. This is the first boat from the *Housatonic* arriving to get help. *Canandaigua's* skipper, Captain John Green, immediately takes the survivors aboard, slips his own anchor and moves at best possible speed for the *Housatonic's* last known position. He arrives there at 2132 and holds off about eight hundred feet. Putting his boats in the water, he rescues twenty-one officers and one hundred and twenty-nine men. A bit after midnight he gets a message to Commodore Rowan aboard *New Ironsides*, several miles further south on the blockade line, to let him know what has happened. In the meantime, Green has transferred fifty-seven of *Housatonic's* crew to the frigate USS *Wabash*.

USS Canandaigua Moved In To Rescue The Crew of Housatonic.
Source: U.S. Navy Historical Center

Now – let us pause for a moment to ask this question: Roughly two miles away, one of the most powerful and modern ships in the United States Navy has been sunk following a catastrophic magazine explosion that didn't quite blow her in two. She was lit, she was stationary, visibility was excellent, yet she sank within sight of lookouts across the blockade line – and *not one saw her go down or noticed anything amiss*. And on top of that, no one ashore – including sentries at Battery Marshall or Forts Moultrie and Sumter – saw or heard *anything*.

Why not?

There are several possible explanations for what happened, and some of them are none too complimentary to the men of the US Navy or the CS Army. First, the explosion was underwater, with the *Housatonic's* hull between the detonation and the *Canandaigua*, the wooden hulled line, and anything south of the *Housatonic*. That at least explains a failure to spot it visually, but it doesn't explain why the ironclads north of *Housatonic* or the sentries at Marshall (who should have been specifically looking for something), Moultrie, or Sumter.

There were some scattered pockets of mist low on the water – we know that from the Galena letter – and it is possible that one or more was between the shore stations and the *Housatonic*, but that doesn't explain why the other ships didn't see it, especially as even the final report categorically states that the weather was clear.

There is the possibility - and this is a strong, realistic one - that the CSA sentries ashore simply took any flash they saw to be a calcium light or incoming artillery round. Those were things that they were used to seeing every day and every night, and human nature being what it is after a while one sees what one expects to see.

There is also the odd fact that on occasion, massive explosions have gone completely unheard by those closest to them. For instance, in 1941 the British battlecruiser HMS *Hood* is destroyed by a catastrophic explosion of one of her magazines while dueling with the German battleship *Bismarck* – and not only her three survivors but witnesses aboard HMS *Prince of Wales* state that they never *heard* an explosion.

Another explanation – and this is almost surely applicable in some cases – is that the sentries simply weren't paying attention. Ashore it would have been tempting and easy to duck behind a revetment or wall to stay out of the wind – after all, the Yankees weren't invading tonight, were they? 'Course not, and if they did, there'd be plenty of warning. And afloat, quite frankly, there was not only the motivation to stay warm but the fact that not every skipper in the fleet was as thorough as Captain

The Long Patrol

Pickering was – some ships may not have had more than one or two lookouts posted, and if they were less than motivated they may very well have missed it.

One might think that the sound of that much ammunition letting go should have been noticed, but in this case I believe we can be a bit more understanding. Most of the explosion went down and into the water, so it's likely that no one except the crews of the *Housatonic* and the *Hunley* actually heard it, and in any event as we've said, it's possible not to hear it at all.

But in any event, the *Canandaigua* has arrived at the wreck site by 2135, making the trip in just about fifteen minutes. *Housatonic's* boats and those of *Canandaigua* are shuttling back and forth, taking the survivors out of the rigging and back to the warmth and safety of the ship now sitting about eight hundred feet away. In the finest traditions of the US Navy, the men more seriously hurt went first and the rest sat quietly in the rigging waiting their turn. One of them was Seaman Robert Fleming, perched safely in the forward rigging and apparently quite dry and comfortable while he waited for his trip across to the *Canandaigua*, this is what he saw:

"...When the Canandaigua got astern, and was lying athwart of the Housatonic, about four ship lengths off, while I was in the fore rigging, I saw a blue light on the water just ahead of the Canandaigua, and on the starboard quarter of the Housatonic."[64]

A blue light. The agreed signal from the *Hunley* to Battery Marshall, and back from the Battery to the *Hunley*. Pay close attention to the time: it's now fifty minutes after the attack, and someone is giving or answering the signal indicating a successful attack.

The question here now is which light Seaman Fleming is seeing: the *Hunley's* or Battery Marshall's? We believe both were given:

"...The signals agreed upon to be given in case the boat

---

[64] Ragan, pg. 181

CSS H.L. Hunley

*wished a light to be exposed at this post as a guide for its return were **observed and answered**...*"[65]

LTC O.M. Dantzler, CSA
Officer Commanding,
Battery Marshall

Mark Ragan asks whether or not Seaman Fleming saw *Hunley's* signal. That is indeed what happened, but it wasn't as close as Fleming thought. According to Fleming's testimony, this is what the scene should have looked like:

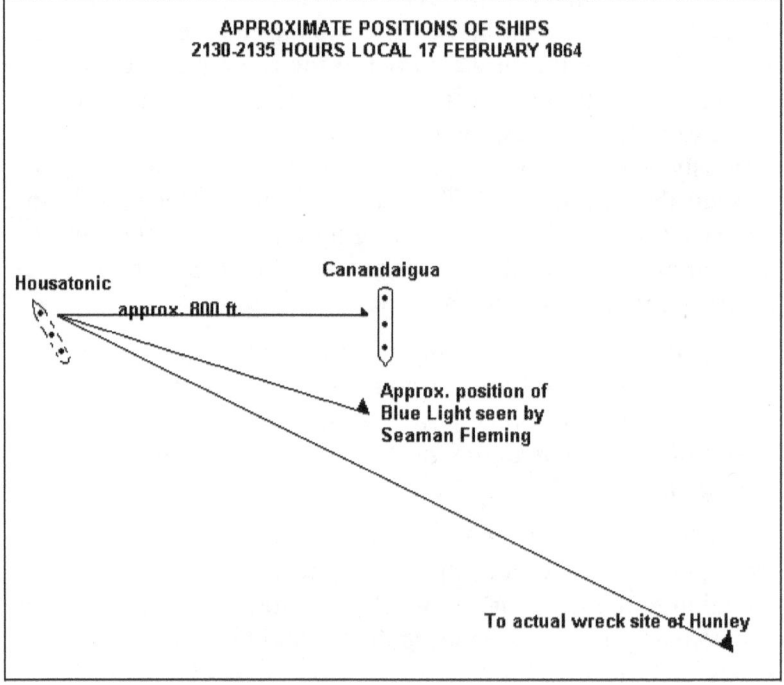

It's important to keep in mind what Seaman Fleming is saying: He is seeing a blue light *forward* of the *Canandaigua*, and the *Canandaigua* is 'lying' – not moving, (something I believe eliminates the possibility that she ran the *Hunley* down,

---

[65] *Official Records*, Series I, Vol. 15, pg. 233

something we'll look at later) off *Housatonic's* starboard quarter, i.e.; the right rear of the ship. In addition, *Hunley* was found about a thousand yards – three thousand feet, or about half a nautical mile – to the east of the *Housatonic*. If what Seaman Fleming saw was the *Hunley,* how did Dixon manage to miss the *Canandaigua,* roughly twenty-two hundred feet away? Why would he have flashed that blue lantern and risked being spotted?

The most reasonable explanation I have been able to come to is, quite simply, that Dixon had no idea the *Canandaigua* was there. Once Captain Green realized the *Housatonic* had been sunk by a torpedo boat, he would have taken every reasonable precaution to make sure that if there were any others out there, he wouldn't be next – and one of those precautions would have been to darken ship. With the moon as bright as it was and a clear sky, he could see clearly enough, and the lookouts would have been most vigilant for anything else that suddenly appeared.

In addition, it's possible that *Canandaigua* may have been hull down to Dixon – that is, anyone looking at her from *Hunley* was looking directly at her end on. With the ship's lights out, sitting half a mile away, and *Hunley* ballasted down so that Dixon's head was perhaps at most sixteen to twenty-four inches above the water, he may very well have never even seen her.

Where exactly has *Hunley* been for the last forty-five minutes or so? As discussed earlier, she was underwater at first, very slowly trying to crank out of the area. However, she wouldn't have been making much headway in the effort – the ebb tide is still running and will be until just after 2200, which means she is trying to work against the current. Dixon probably got her turned about quickly enough after the initial attack, but they're working against a current that is almost as fast as their maximum possible speed this night.

Clive Cussler feels that Dixon took her out of the area under power, but I respectfully disagree. The effort would simply be too much for her battered crew, so Dixon decides to just let her drift. She is moving with the tide very nearly as fast as she would have been under power in any event, and in forty-five minutes this puts her some distance east of the *Housatonic* – exactly where Cussler's team found her in 1995 when he realized that *Hunley* could not have fought the tide all the way back to Breach Inlet. If they weren't cranking – pretty much a given – they should be able to hold out for a while, but they will be coming up on the end of their endurance. Remember that we know that with no effort at all, they could hold out for two hours – and even if things were getting unpleasant, the threat of sudden, violent death if discovered would have been an impressive motivator to hold out as long as they possibly could.

But regardless of what they were trying to do, by 2130 the crew would have been approaching the end of their rope. They are tired, scared (or at best very nervous), approaching hypothermia, likely dehydrated and suffering from extremely high levels of carbon dioxide in the hull and the first symptoms of anoxia. If they were suffering from shock-induced internal injuries, loss of blood may have made things even worse.

Could they have ridden it out using the snorkel? It is indeed possible – a great deal of work went into the design and installation of the snorkel, and as mentioned earlier, the design of the sub suggests that one of the crew members may have been specifically assigned to operate it. But it appears that the snorkel never worked satisfactorily, if it worked at all. Will Alexander suggests that it was an iffy proposition at best, and that opening the hatches was the most efficient way to get fresh air into the hull. The snorkel may have at best only been a quick stopgap, able to get a few quick breaths to the crew, but nowhere near sufficient to keep them breathing for long. A quick surfacing, not only to signal Battery Marshall but also to get air into the hull, is now a life-or-death necessity and Dixon knows it.

What I've written so far can reasonably come under the heading of informed speculation. We've got eyewitness accounts, we've got solid data, and we can extrapolate from there – something I believe I've done well enough to make a plausible argument that brings us to this point. From here it's theoretical, but I believe it is a theory that explains how nine brave men found themselves trapped on the bottom of the Atlantic Ocean.

The Long Patrol

At around 2130, Dixon and his men are almost out of air and definitely out of options. Worse still, the *Hunley* is slowly bleeding to death from sprung seams in the forward and aft ballast tanks plus the aft stuffing boxes. This, by the way, is where Lieutenant Alexander's loss truly becomes critical – had he been in his seat aft with his thorough knowledge of the boat and how it worked, he might have been able to realize that something was very wrong, just from the way she handled. That, in turn, would have given them a chance – a thin one – at survival if they had abandoned ship.

But Alexander is hundreds of miles away in Mobile, unable to help. Dixon is in command, and he has to make the call. That he does: take her up. The crew – hypothermic, anoxic, and exhausted – probably doesn't even realize that the *Hunley* is handling strangely. Dixon makes sure they surface just enough to get the hatch open for him to show the light.

All of this, by the way, is solid evidence that the *Hunley* had not suffered any damage to her hull caused by direct fire of any kind during the attack. Had she been taking on water through any of the holes later found in her hull, she could not survived underwater for more than a few minutes – and for that matter, she could not have remained afloat much longer even on the surface, and we know she wasn't. Dixon lights the lantern *before* opening the hatch – to do otherwise would call attention to himself through the flare of the match, and it's going to be difficult enough just to get the lantern up without calling down the Furies upon them.

Dixon pops up through the hatch and instinctively scans the area.

The only known image of Col. Olin M Dantzler. Col. Dantzler was killed leading a charge to take Fort Dutton on June 2, 1864. Source: Official Records of the Union and Confederate Armies

CSS H.L. Hunley

Hunley is pointed more or less towards Battery Marshall, while *Canandaigua* is actually ahead of him off his starboard bow, though exactly how far we'll never know. Dixon points the lantern at Battery Marshall and quietly hopes that he can get the message to the fort before he's spotted. He's putting a great deal of trust in the watchfulness of Moultrie's sentries, and we already have a sneaking suspicion it may be misplaced.

We don't know how long he showed the light for, but it had to be at least for a minute or so. Dixon cannot see the return signal from Fort Moultrie – remember, he is far too close to the water to see anything of the fort's ramparts roughly four miles away. After about a minute, Dixon closes the hatch – he can't light the candle before doing so, because that could increase their chances of discovery – and gives the order to take her down. Some fresh air has gotten into the hull, but not enough to alleviate the conditions inside.

Hunley is still taking on water from the leaking seams, and having the tanks filled one more time is the final straw. As she slides under, Dixon may probably make Horace Hunley's last mistake – with the boat far heavier than she should be and with her buoyancy delicate under the best of circumstances, Dixon may be trying to light the candle as she moves rather than having her remain still until he can get some illumination and check the depth gauge.

As he tries to get it lit in the darkness, *Hunley* is moving forward – and down – at about two knots. At that speed and moving at a gentle enough angle that the numbed crew members will not realize anything is wrong until it is too late, it will take just a few moments or so for her to slam hard into the bottom, the base of her prow grazing first then her keel digging in as she skids to a stop, lying on her starboard side.

As *Hunley* grinds to a halt, she is now lying on her starboard side at roughly a 45° angle. Water comes pouring over the ballast tank bulkheads and along the starboard side of the hull. Increased water pressure at that depth pushes more water through the seams into the *Hunley*. Even if the ballast control valve is still on its post, the flooding through the seams is unstoppable. The crew, seated along the port side, is in complete darkness and probably draped over the crank but still more or less at their stations.

The water is *cold* – the average water temperature in Charleston in February is 50° and a healthy man suddenly immersed in that water can survive for as long as two hours in normal street clothes. But these men are not healthy, they are wearing clothes that could be described as threadbare and may even be barefoot, and they are already badly chilled. In addition, the water temperature thirty feet down is several degrees lower than at the surface. Their survival time is going to be measured in minutes.

Their instinct will be to either start pumping the tanks out or try to crank their way out of it. Neither one will save them. With the *Hunley* firmly against the bottom and mostly upright, the ballast pump intakes are blocked by sand and mud while increasing amounts of water in the hull will prevent the crew from getting her up off the ocean floor no matter how hard they crank – and, not incidentally, using up their oxygen like a modern fighter plane in full afterburner uses its fuel.

I suggest that Lieutenant Dixon, still at his post forward, went first. Possibly stunned or otherwise incapacitated by the impact, he is down against the forward bulkhead and immersed by the icy water. His body temperature, already low, sinks rapidly. Even if he wasn't already unconscious, he will be disoriented, unable to reason or function properly. George Dixon will slowly fade out in the next few minutes, and then pass over to the other side.

His crew – increasingly hypothermic, disoriented and confused – will soon follow. Given their likely condition before impact with the bottom, they are already badly behind the curve when it comes to dealing with an emergency. Now soaked in near-freezing water, with low air levels, steadily rising water and in utter darkness with their commander unable to help, they cannot possibly react quickly or precisely enough to even buy themselves a few more minutes.

CSS H.L. Hunley

Were there any other possible causes for the loss of the Hunley?

The scenario I outlined above is far from the *only* possibility – it is *a* possibility, and I believe it to be a strong one that fits the facts as we know them. But over the years, other scenarios – perfectly logical and reasonable – have been forwarded, and they are worth looking at

Was the Hunley struck by another ship?

This scenario has been brought up several times, most notably by Clive Cussler. The various takes boil down to this: *Hunley*, either on the surface or just below it, is hit by one of the vessels moving to *Housatonic's* aid. She is either run directly into the sea floor or, rolling out of control with a gash in her side, floods and sinks moments later. Like all of the alternate scenarios it cannot be completely ruled out, but I believe the evidence we have works against it.

There were only three vessels in the immediate vicinity of the attack: *Housatonic,* of course, but she is immobilized and sinking, incapable of doing anything against the *Hunley.* The *Hunley* itself is moving away to the ESE on the outgoing tide. The only other ship that we know entered the area is USS *Canandaigua*, moored about a mile and a half away until about thirty minutes after the attack. Could she have collided with the *Hunley*, especially in light of the possibility that Lt. Dixon didn't know she was there?[66]

Seaman Fleming – whose perch in *Housatonic's* forward rigging enabled him to see the blue light – says this about *Canandaigua*:

"When the Canandaigua got astern and was lying athwart..."

Remember, Fleming says, 'lying' – that is, not moving – when he sees the blue light. The *Canandaigua* would have had to come to a halt to launch and recover her boats for the rescue effort. This in and of itself would seem to eliminate her as

---

[66] See pg. 177 - MJK

The Long Patrol

Hunley's executioner, but there are other details to take into account as well. Again, Seaman Fleming speaks:

> *"...And was lying athwart the Housatonic, about four ship lengths off, I saw a blue light on the water just ahead of the Canandaigua and on the starboard quarter of the Housatonic..."*

Take a look back at the drawing on page 176 – four ship lengths (Depending on the ship - *Canandaigua* was slightly longer than *Housatonic,* 228 feet vice 207) put her between 828 and 912 feet away. But for the sake of argument, let's assume that Fleming underestimated the distance and that *Canandaigua* was a full thousand feet away at 2135. We know though that *Hunley* had drifted approximately *three thousand* feet ESE from *Housatonic* by the time she showed her light. Seaman Fleming could very well have misjudged where the light was, but it's unlikely that he misjudged his distance from the *Canandaigua* - in clear conditions – by a factor of three.

There are also no clear signs on the *Hunley* itself of a collision. There is damage to the hull both fore and aft, but none of it is consistent with being struck by fourteen hundred tons of warship moving at ten knots – had she struck *Hunley*, it seems reasonable that the damage would have been far greater.

Is it possible that a fourth vessel entered the area? Possible, yes, but not likely. *Housatonic's* boats made directly for the nearest ship they saw – *Canandaigua.* The official reports – from Captain Green on *Canandaigua*, Lieutenant Higginson from *Housatonic,* Admiral Dahlgren on the *Philadelphia*, and the final Board of Inquiry report – do not mention any other vessels coming to the aid of the stricken warship. The only other vessels even mentioned are USS *New Ironsides*, whose skipper was the senior officer present on the blockade line that night and wasn't notified until well after midnight, and USS *Wabash*, which took aboard 57 of *Housatonic's* crew – but again, not until after midnight.

By that time, there's every reason to believe that Dixon and his crew had already been dead or dying for at least two hours. There's one other drawback to the collision theory - admittedly, it would have been difficult at best for Dixon to see another ship even if navigating on the surface with hatches open, but every ship that could have conceivably come running to *Housatonic's* aid at this point was steam powered. On a clear moonlit night with almost no wind, Dixon should still have been able to first hear, then see and avoid another vessel, even from his limited vantage point.

Was she swamped by another ship or a sudden squall?

The scenario makes sense – *Hunley* sat very low in the water, and all it would have taken was one rogue wave to wash over her. The swamping theory also has an argument from the loss of LT Payne's crew – many witnesses felt that the *Hunley* had been swamped by the ship that was passing her as they came alongside CSS *Etowah*. In any event, this could have happened in one of two ways: the wash from a passing ship or a sudden squall brewing up.

I believe we can safely eliminate a passing ship based on what we know about how many ships were in the area and where they were. No vessel afloat in 1864 could have created a wake that would have swamped *Hunley* from at least two thousand feet away. That leaves the squall, and the evidence of that is slim.

The only mention we have comes from the February 29th edition of Robert Rhett's Charleston *Mercury*. That day, Rhett published the story of the Federal picket boat captured by the *Indian Chief's* crew, and how the prisoners revealed the loss of the *Housatonic*. Rhett proceeded to claim that Dixon and the *Hunley* were safe – and that the *Housatonic* was originally reported as lost in a gale. This has, in turn, led to the question of whether or not *Hunley* was lost like this.

It's not at all impossible that a sudden squall could have boiled up out of nowhere – after all, the Atlantic in February can be fickle and dangerous. But we should look at the accounts of the weather:

Court of Inquiry: *"The weather at the time of the occurrence was clear...wind moderate from the northward and westward, sea smooth...*"[67]

Galena Letter: *"There was but little wind or sea, the sky was cloudless...a slight mist rested on the water, not sufficient, however, to prevent our discerning other vessels on the blockade two or three miles away."*[68]

Remember that *Hunley* is at most about three thousand feet from *Housatonic* – yet none of the cruiser's survivors nor anyone aboard *Canandaigua*, nor anyone else for that matter has mentioned a sudden gust of wind that would have had to have been violent enough to kick up 2' - 3' foot waves directly alongside the rescue operations. Keep in mind also that when Lt. Payne accidentally sent the *Hunley* to the bottom, she had both her hatches open and four of her crew were able to escape. Exhausted as he was, Dixon would have had to simply push upwards to escape as she filled – but he was at his post when she went down, as was everyone else. There was no apparent attempt to escape, as may have been expected. In addition, for *Hunley* to have been swamped would again require Dixon to completely miss a vessel bearing down on him – and even as tired and cold and ill as he may have been, that seems difficult to accept.

Could the Hunley have escaped?

I believe that the most important variable here is the detonation of the *Housatonic's* magazine at approximately 2047. If one assumes everything else remains the same – most notably the premature firing of the torpedo, but the magazine does not cook off – *Hunley* will be badly rattled, though not fatally damaged.

But...

If the magazine doesn't explode, *Housatonic* is a hurt but still very formidable adversary. She does have a good-sized hole in her -- the injury is probably not terminal, and her well-trained crew is moving quickly to secure the damage. On the other

---

[67] *Official Records*, Series II, Vol. 1, pgs. 332-333

[68] See Appendix 4.

CSS H.L. Hunley

hand, she is moving and as she moves, her main battery can be brought to bear on the *Hunley*. The boat, of course is trying to get out of the way as quickly as possible – but remember that gun Ensign Craven and his crew were preparing to fire. Once they recover from the jolt of the torpedo's detonation, they will be back in the fight with a vengeance, as will every other gun that can be brought to bear.

Housatonic is capable of destroying much larger and more powerful adversaries at a far greater distance – even with *Hunley* moving, the skilled, competent gunners will be able to track and hit the little sub without any serious difficulty. Remember that *Housatonic's* gunners are already at their stations and preparing to fire. If the ship is not sinking, it will be striking back at its assailant. Smaller cannonballs might have not done too much damage to the *Hunley*, but the massive 24-pounders and the 11' Dahlgren are capable of swatting it literally out of the water. But let's assume that *Hunley* has evaded the fire that *Housatonic* is throwing at it and manages to get clear of what modern submariners call the 'flaming datum'. Dixon will still quickly realize that the ebb tide is keeping them from making any real headway.

What does he do? Does he decide to simply drift and wait, as he appears to have done in history? It may very well have been his only realistic option, but it puts him in the middle of a hornet's nest – once *Housatonic* starts firing, the rest of the line will start converging on her, even if no one else heard or saw the torpedo attack. Missing the initial attack can and has been explained, but there will be no way to miss the thunder of continued heavy gunfire. Dixon can try to wait it out, perhaps running on the snorkel as best he can and hoping he can eventually try and creep out of the area after about 2200. He then runs a real risk of being run down by another ship as they arrive to help *Housatonic*, but if *Hunley* can survive that long, she has a chance.

It is far from a sure thing – and if he cannot get out of the area, he is going to be inching closer and closer to the 2400 – 0100 'point of no return' that would end up putting her back at Breach Inlet near or after sunrise. At that point, with the crew possibly exhausted beyond recovery, Dixon's only realistic choice may have been to surface and surrender. They could surface and drift out of the area, but there is no guarantee which way they will go. To summarize – *if* there is no magazine explosion, *if* the *Housatonic* does not blow her out of the water, *if* they can avoid detection or ramming by the other ships on the line, and *if* the crew can hold on long enough to do that, *Hunley may* be able to get back - but it will still be a very dicey proposition at best.

So, to summarize:

Hunley leaves Battery Marshall at 1900 hours. Just over ninety minutes later, she has gotten into attack position on *Housatonic*. Dixon executes a slow, methodical approach that is discovered at the last second and goes in for the kill. *Housatonic's* crew, alerted, tries to get the ship underway but inadvertently brings the ship's magazine directly into the line of fire.

Upon impact with the side of *Housatonic's* hull, *Hunley's* torpedo – suffering from a series of relatively minor individual problems but in total catastrophic ones – partially disintegrates or deforms, while the impact ruins the triggering lanyard mechanism and/or the torpedo's fuze itself. As *Hunley* backs away the torpedo fires prematurely, setting off *Housatonic's* magazine in a sympathetic detonation. The explosion's shockwave springs *Hunley's* seams fore and aft, starting a slow and unstoppable flooding primarily through the aft ballast tank.

Unable to fight the ebb tide still running, Dixon elects to let *Hunley* drift submerged for about forty-five minutes. This results in more flooding and causes the crew's reflexes and abilities to be further degraded by lack of oxygen. Dixon surfaces just long enough to flash a blue lantern towards Battery Marshall then dives once more. However, by this time *Hunley* has taken on enough water that her crew, impaired from anoxia and cold, cannot realize that their ship is sinking. Probably believing themselves to be straight and level, and without a lit candle to see their instruments, they sail *Hunley* directly into the ocean floor. Jammed tight and with water flooding in, they die from hypothermia and drowning within the next twenty minutes. Not precisely *Quod erat demondstrondum*, but I believe it to be the most reasonable, logical and most importantly, likely scenario.

Whatever happened, it was almost certainly over and done with by around 2135. By 2140 they were certainly incapacitated and unconscious. As the cold water slowly filled the hull, their hearts slowly ran down and stopped, probably around 2200. I believe this also excludes the possibility (again, posited in the film) that the crew, facing a slow, agonizing death, opened the ballast valves and flooded the boat – these men died from a combination of hypothermia and anoxia.

Water will continue to seep into the hull but it will not fill it anytime soon. In a chilling discovery after the salvaged hull was entered, archaeologists found stalactites formed from years of water slowly, steadily dripping through leaking seams.[69] Since stalactites cannot form in water that meant there had to be fairly large voids inside *Hunley*.

For a very, very long time.

Tradition in the United States Navy is that submarines and submariners lost in the line of duty are said to be "still on patrol". Dozens of submarines from the Second World War are still said to be on that lonely, eternal mission.

The attack submarine USS *Thresher* (SSN-593) has been on patrol off the coast of Massachusetts since April 10th, 1963, betrayed by a nearly microscopic broken weld somewhere in the maze of piping that threaded through her hull.

Her older cousin *Scorpion* (SSN-589) is still on patrol in the blue waters of the Atlantic, four hundred miles southwest of the Azores. Her executioner is unknown to this day, with possible claimants as varied as a Soviet submarine, or one of her own torpedoes gone berserk inside the boat's working spaces.

---

[69] Brian Hicks and Schuyler Kropf, *Raising The Hunley: The Remarkable History And Recovery Of The Lost Confederate Submarine*. Ballantine, New York, 2002, pg. 218.

The Long Patrol

We do not know with certainty why the *Hunley* sank, but sink she did, with her crew, just about two hours after leaving her dock and changing naval warfare forever.

Her long patrol has one hundred and thirty-one years to go.

CSS H.L. Hunley

VI: "...THEM SAFE FROM PERIL IN THE DEEP."

It was *two days* before anyone at General Beauregard's headquarters realized the *Hunley* was missing. As mentioned earlier, no one ashore saw or heard the *Housatonic* get hit. Strange, but ultimately believable through a series of unusual physical effects connected with the explosion and the results of old-fashioned human nature. What is however more difficult to believe is that after the troops at Battery Marshall acknowledged the *Hunley's* signal, no one thought it odd that the boat never showed up, and worse still never reported it. This definitely bears looking into for a moment, and that's probably a good idea – because it's at this point in the narrative that LTC Dantzler's message has to be looked at.

The entire body of the message was as follows:

"*Head Quarters, Battery Marshall*
Sullivan's Island
Feb. 19th, 1864

"*Lieutenant –*

I have the honor to report that the torpedo boat stationed at this post went out on the night of the 17th instant and has not yet returned. The signals agreed upon to be given in case the boat wished a guide for its return were observed and answered. An earlier report would have been made of this matter, but the Officer of the Day for yesterday was under the impression that the boat had returned, and so informed me. As soon as I became apprised of the fact, I sent a telegram to Captain Nance assistant adjutant-general (On General Beauregard's staff – MJK), *notifying him of it.*

"*Very Respectfully,*
Col. Dantzler"

The Long Patrol

The first question one tends to ask in these untrusting days is, "Can we believe LTC Dantzler? Did anyone actually see and answer the blue light at all?" Mark Ragan says yes – for his work, the Dantzler message is a critical facet of the evidence, proving that *Hunley* survived the initial attack. Clive Cussler – the man whose work found the *Hunley* – strongly suggests that Dantzler's memo is not trustworthy, and that the Colonel may have been trying to cover up a less than stellar performance on his part. So - what happened at Battery Marshall after 2130 on February 17th?

My feeling is that both men are right, but it is admittedly difficult to prove. First, we only have LTC Dantzler's word that the blue light was seen and answered. A statement from the unnamed OOD or anyone else on the battery would be very helpful, but it's not likely at this late date that one will ever appear. On the other hand, if Dantzler had pressured the OOD or anyone else at the Battery to falsely claim that the light had been seen and answered, it would not have taken long for that to make it back to General T.S. Ripley, the CSA commander on Sullivan's Island (which included Fort Moultrie and Battery Marshall) or to General Beauregard himself. The Sullivan's fortifications – Moultrie and Batteries Bee, Beauregard, and Marshall – were small, as was the number of men stationed there. Everybody would have known everybody else, and if there had been a cover-up of any kind, it would have been known and known quickly.

Whatever Beauregard's military drawbacks, his personal integrity is absolutely unquestioned. Had he even suspected Dantzler of such a thing, it would have meant a swift transfer from the relatively safe posting at Marshall to someplace far more dangerous, or being busted out of the service altogether. There is another thing to keep in mind as well – Mark Ragan's research found evidence that Dantzler and Dixon personally discussed the matter of the signals sometime during the day on the 17th.[70] We have no indication that Dantzler was anything other than a competent, conscientious officer, and if he gave his

---

[70] Ragan, pg. 171

CSS H.L. Hunley

word to Dixon – a fellow officer – the standards of the time demanded that he keep it. Now, having said all that, I believe that here human nature rears its head once again. Dantzler certainly knew the importance that the CSA command in Charleston put on the program, and he did nothing that directly contributed to the loss of the *Hunley* – but there were going to be unpleasant consequences once it got out that the boat was missing almost two days after its departure. He's got to do something, but it would not be human were he not to try and minimize the blame a little.

Given the way this sort of thing happens in the military, the sequence of events was probably like this: The OOD on the 18th was told that the blue light had been spotted, and the reply given. (The difficulty here is that no one aboard *Canandaigua* or any of the other ships saw a light coming from Battery Marshall. That doesn't mean it didn't happen, but it does plant the seed of doubt.) The OOD asks, "Did the torpedo boat return?" He probably received an answer along the lines of, "They flashed their signal at about nine-thirty, and the guide boat was supposed to go out after them" – because whoever told him that knew that was what was supposed to happen. The OOD, for whatever reason, never followed up on it and then told LTC Dantzler that the *Hunley* had been brought back in by the guide boat.

Dantzler, commanding a coastal defense battery within sight of an enemy fleet, has other things to do on the 18th than follow up after the *Hunley's* crew – Charleston and the fortifications are under fire again, and heavily so - and doesn't get a chance to stop by the Breach Inlet dock or delegate someone else to do it until the morning of the 19th, then finds the *Hunley* gone. Some quick questions of the sentries raise disturbing possibilities, so he goes back to Battery Marshall, asks more questions, and realizes that the boat has been missing now for roughly thirty-six hours, hence the fast telegram to Captain Nance at Beauregard's HQ and the follow-up memo for the records.

My feeling that at worst, Dantzler stands guilty of not following up sooner – given the fact that they are under fire again early on the morning of the 18th, it would seem to be a forgivable mistake. That would seem to put it all into perspective – an accurate (if not entirely forthcoming) telling of what happened the night before, followed by the indirect suggestion that it wasn't his fault, and knowing that with incoming fire no one would be asking questions anytime soon.

On the morning of the 20th, the sentries ashore finally saw something – three masts poking out of the water, surrounded by a small flotilla of vessels, not quite three miles out from Breach Inlet. Dantzler knew about this immediately, and another message ran down the road to Fort Moultrie, then across the harbor to General Beauregard. Although there aren't any other communications to back it up, it's likely that if a full-dress investigation wasn't underway as to the whereabouts of the *Hunley* and whatever connection it may have to the apparent wreck offshore, there certainly is now. Likely everyone on the sentry rosters and manning the guns for the night of February 17th-18th is being spoken to in order to get some idea as to what's going on.

In the meantime, Admiral Dahlgren has been notified late on the 18th that one of his ships is gone. Lieutenant Higginson has been sent out to Port Royal, SC, Dahlgren's present location, to brief him on the events of the 17th. Captain Pickering is recovering nicely, but he's not able to travel. Higginson gives Dahlgren the bare-bones facts of the matter, and praises the fast and gallant assistance of Captain Green and the *Canandaigua*. Some crewmembers have been transferred to the frigate USS *Wabash*, still on the line. *Canandaigua* herself is still back at Charleston, and she's one of the ships now anchored around *Housatonic's* wreck.

A diver went down to the blasted hulk on the 20th to make a preliminary survey – apparently there was some thought at first that the ship might be refloated, but Captain Green quickly disabused them of that notion:

"...*Her spar deck* (is) *about 15 feet below the surface of the water. The after part of her spar deck appears to have been entirely blown off. Her guns, etc., on the spar deck, and probably a good many articles below deck, can, in my opinion, be recovered by the employment for the purpose of the derrick boat and divers.*"[71]

The reader perusing the actual report will note one other detail, the dog that does not bark in the night: There is no mention whatsoever of *Hunley's* wreck being found entangled within or alongside *Housatonic's* hulk – but more upon that subject later.

Dahlgren's official response to all this is firm, direct, and professional, as will be seen below. His personal response was probably volcanic – after all the warnings and standing orders, one of his most powerful ships was blown out of the water while at anchor. But it could not be undone, and after the initial shock and anger, Dahlgren got to work on the 20th and sent a message to Secretary of the Navy Gideon Welles:

"...The Department will readily perceive the consequences likely to result from this event; the whole line of blockade will be infested with these cheap, convenient, and formidable defenses, and we must guard every point. The measures for prevention may not be so obvious.

"I would therefore request that a number of torpedo boats be made and sent here with dispatch; length about 40 feet, diameter amidships 5 to 6 feet, and tapering to a point at each end; small engine and propeller, an opening of about 15 feet above with a hatch coaming, to float not more than 18 inches above water, somewhat as thus sketched..."[72]

If this sounds familiar, it should. It is very much a description of CSS *David*, which makes sense – Dahlgren was firmly under the impression that a *David* had killed the *Housatonic*:

"SIR: I much regret to inform the Department that the U.S.S. Housatonic, on the blockade off Charleston, S.C., was torpedoed by a rebel "David" and sunk on the night of the 17th February at about 9 o'clock...

---

[71] *Official Records*, Series 1, Vol. 15, pg. 331.

[72] *Official Records*, Series 2, Vol. I, pgs. 329-330.

The Long Patrol

"From the time the "David" was seen until the vessel was on the bottom a very brief period must have elapsed; so far as the executive officer (Lieutenant Higginson) can judge, and he is the only officer of the Housatonic whom I have seen, it did not exceed five or seven minutes."

One wonders what Dahlgren would have done with a couple of squadrons of Federal *Davids*, capable of making ten to twelve knots each. The thought of them charging into Charleston Harbor to sink anything at anchor, or going after runners, leads to some interesting speculation on what naval combat may have evolved into. Then, after advising Secretary Welles that he was ordering his own floating torpedoes and reinforcing the countermeasures already taken, Dahlgren makes one other suggestion, one that must have set hearts pounding throughout the blockade line off Charleston:

"...I desire to suggest to the Department the policy of offering a large reward of prize money for the capture or destruction of a "David"; I should say not less than $20,000 or $30,000 for each. They are worth more than that to us."

Secretary Welles never did have to write a check. There was only one more *David* attack, against the USS *Memphis* in the Stono River on March 4th. The attack was perfect, but the torpedo failed to explode, apparently due to a failure of the acid-based trigger system. *David* tried again on May 1st, but turned back for reasons that are unclear. The only *Davids* ever captured were that handful that lay beached like dead sharks when Charleston was taken in 1865, though it is possible *David* herself was the one that ended up in Annapolis, MD.

Be that as it may, the suggestion of financial reward does not seem to have encouraged aggressiveness in the ships of the blockade line: Captain John Rowan of USS *New Ironsides* advised Dahlgren on February 23rd that only one of his picket boats – the smaller sailing craft that were used as lookouts and patrol ships – had even shown up for duty since the *Housatonic* was lost.[73]

---

[73] Ragan, pg. 188.

CSS H.L. Hunley

While Admiral Dahlgren was motivating the men of the South Atlantic Blockade Squadron, General Beauregard and his staff were still trying to determine what happened to the *Hunley*. A few accounts – sadly untrue - had her putting in at other locations along the coast, and when combined with the delay at Battery Marshall in realizing she was missing, it becomes possible that on at least one occasion and maybe more *Hunley* actually did end up drifting some distance from Charleston before being recovered. This, though, is one matter that we will never be able to nail down with any certainty.

The final answer, though, came late on the night of February 26th in a most unexpected fashion. A Federal picket boat (hopefully not the only one that had shown up) apparently got its bearings crossed and passed dangerously close to Fort Sumter. As it turned out, a small boat assigned to the *Indian Chief* was on patrol in the area and leapt into action. Within minutes, the CSN patrol had cut off the Federal picket and took it and its crew prisoner.

Interrogations in those days were not the thoroughly unpleasant things we know them to be today, and people talked with surprising ease. What these men told their CSN interrogators confirmed the best and the worst of what had been suspected since February 17th: USS *Housatonic* had, indeed, been sunk by a submarine torpedo boat. The boat had not been captured, nor had it been sighted again since. Beauregard immediately sent a message to CSA Headquarters in Richmond, making sure they knew what had happened, and that there was little hope for Lieutenant Dixon and his men.

Little hope, indeed. Beauregard and Ripley had probably scoured the coast and sent messages everywhere they could think of, trying to find their lost men, and there had been nothing. Word had gotten out to the local newspapers – the *Mercury* and the *Courier* – about the sinking of the *Housatonic,* but they were unfailingly (and vaguely) cheerful about the health and safety of the crew, and utterly without clue as to their actual fate – all the more painful because some poor soul at Beauregard's headquarters would have been detailed to run down each of those reports only to discover their untruth.

But at the end of the day the search was finally called off. By February 27th, Beauregard, Ripley and Dantzler had probably gathered to quietly review everything they knew – the reports from the prisoners, the queries sent to every hole-in-the-wall port or beach with a ten-cent dock up and down the coast, and the cruel, bitter arithmetic of just how long the *Hunley* and her crew could have survived out there. Beauregard would have asked for conclusions, but it was really just a formality. Lieutenant George Dixon, CSA, his crew, and the submarine torpedo boat *H.L. Hunley* were declared missing and presumed lost. Military tradition requires a brief, direct announcement, in this case from Beauregard's HQ to Richmond:

"*HEADQUARTERS, ETC. March 10, 1864.*

"*SIR: I am directed by the commanding general to inform you that it was the torpedo boat H.L. Hunley that destroyed the Federal man-of-war Housatonic, and that Lieutenant Dixon commanded the expedition, but I regret to say that nothing has since been heard of Lieutenant Dixon or the torpedo boat. It is therefore feared that that gallant officer and his brave companions have perished.*

"*Respectfully, your obedient servant,*
"*H.W. FEILDEN,*
Captain and Assistant Adjutant-General."[74]

Dulce et decorum pro Patria mori. There would be one last official message, officially sent in reply to a query by General Dabney Maury at his lonely post in Mobile, and certainly forwarded to Will Alexander:

"*OFFICE SUBMARINE DEFENSES,*
Charleston, S.C., April 29, 1864.

"GENERAL: *In answer to a communication of yours, received through headquarters, relative to Lieutenant Dixon and crew, I beg leave to state that I was not informed as to the service in which Lieutenant Dixon was engaged or under what orders he was acting. I am informed that he requested*

---

[74] *Official Records,* Series II, Vol. I, pg 337

CSS H.L. Hunley

Commodore Tucker to furnish him some men, which he did. Their names are as follows, viz: Arnold Becker, C. Simkins, James A. Wicks, F. Collins, and Ridgeway, all of the Navy, and Corporal C.F. Carlsen, of Captain Wagener's company of artillery.

"The United States sloop of war was attacked and destroyed on the night of the 17th of February. Since that time no information has been received of either the boat or crew. I am of the opinion that the torpedoes being placed at the bow of the boat, she went into the hole made in the Housatonic by explosion of torpedoes and did not have sufficient power to back out, consequently sunk with her.

"I have the honor to be, General, very respectfully, your obedient servant,

M.M.GRAY,
Captain in Charge of Torpedoes"[75]

The film *The Hunley* has a touching and poignant scene at its conclusion where Beauregard attends a memorial service for his submariners, but it 's not clear if there was ever any kind of official ceremony, and the public was never formally told of the loss of the boat. It would have been a blow to morale, especially as despite the loss Federal forces were bombarding Charleston as vigorously as ever. In addition, Beauregard knew quite well that there were Federal intelligence agents in Charleston, not to mention his own men who might be captured or become deserters. He had no intention of letting the Yankees know that his men hadn't made it back, and he was perfectly happy to let them think that the *Hunley* might still be out there, quietly stalking its next victim. Every dollar they had to spend on

---

[75] *Official Records*, Series II, Vol. 1, pgs 337-338. Captain Gray had a most interesting career with the CSA – he was jailed for six months due to what are simply referred to as 'irregularities' in purchases of rope for the torpedo booms. After that he seems to have had a change of heart and made several attempts to desert to Union forces, finally succeeding some months after *Hunley's* loss. His aid was invaluable in clearing the minefields that by the end of the war littered Charleston harbor – Burton, pg 275

The Long Patrol

countermeasures was a dollar they couldn't spend on ammunition; every man on lookout was one less gunner. It must have torn Beauregard apart to do it, but there would be no formal announcement of *Hunley's* loss, no military remembrance of their sacrifice.

There would be no memorial service at St. Michael's. No black-draped cortege down the long road out to Magnolia, with a steady beat tapped out on muffled drums. No perfectly aligned ranks of riflemen to fire a final volley in salute. No gentle, dignified words from a clergyman offering solace and comfort. No clatter of sword and spur, no glitter of medal, brass and braid. Not even a last Stainless Banner dipped in their honor.

Just cold silent darkness and unyielding black iron that kept nine men frozen forever at 2135 hours, February 17th, in the Year of Our Lord Eighteen Hundred And Sixty Four.

Time passed.

Hunley and her crew – to be blunt – were forgotten with depressing speed. The bombardment of Charleston continued almost without letup, and if anything worsened. The blockade was never 100% effective, but fewer and fewer runners were making it through, and the list of ships that didn't was getting longer: *Prince Albert, Flora, Rattlesnake, Constance.*, and dozens more either taken as prizes, or simply turned into battered splinters of weathered wood and little floating islands of debris.

Ashore it was just as bad and getting worse. The specter of utter and absolute defeat was now starting to loom large, and no amount of cheerful (and completely untrue) reporting by the papers could hide it any longer. Even the few society events that still pressed gamely on couldn't make people forget, especially as the sound of Federal artillery rounds drowned out the minuets and waltzes.

The food supply was getting tighter and tighter, and prices – high since Secession – became astronomical, with even the cost of a barrel of flour passing the $1,000 CSD mark.[76] The thought

---

[76] Charleston wasn't alone in this and in some places, most notably Richmond itself, troops had to be called out to control protests - MJK

CSS H.L. Hunley

of preparing the Holy City for a house-to-house battle was even entertained, with barricades to be placed at critical points throughout the city.

A contemporary impression of the bombardment of Charleston. At this time, deliberately bombarding civilians was seen as the act of scoundrels. Source: Harper's Weekly.

Fortunately, that idea was never implemented, but that may only have been because of another yellow fever epidemic that swept the city in August of 1864. With undernourished people in poor condition with weakened immune systems, the casualty count mounted quickly and stayed there until cooler weather set in towards the beginning of November. Many of the victims were Federal prisoners held at the camps that dotted the Charleston area, but the Plague Time was not discriminating in its touch. Young, old, rich and poor alike fell in its wake, and the long trip out to Magnolia must have become sadly familiar. Officially though, the strategic situation was just fine and nothing was wrong until the morning of December 21st, 1864.

Federal General William Tecumseh Sherman had been 'making Georgia howl' since his capture of Atlanta on September 2nd – and one of his first acts was to order the immediate evacuation of that city's residents, an act of hard military logic that made tens of thousands of Atlantans refugees and which cast a cold, terrifying shadow over the rest of the South. Atlanta had been in the very heart of the Confederacy, and no matter what the papers babbled on about regarding strategy, plans, and grim revenge, the facts were plain: the Confederate Army could no longer defend its homes, the end was now in sight. And the end would have an apocalyptic touch to it that the cheering crowds on the Battery in 1861 could never have foreseen. First, Sherman burned Atlanta on November 15th, a week after Abraham Lincoln had been reelected President. He then moved out with sixty-two thousand men, who had but one simple directive: march to Savannah.

It took Sherman thirty-six days to march two hundred and eighty-five miles, and every inch was a horror of burned farms and homes, slaughtered livestock and stolen property. Sherman had ordered that his men live off the land, and Georgia – so far essentially untouched by war – was bursting with food and provisions, often better than what their own quartermasters had been providing. The rail lines were pried up; the rails heated and twisted into giant loops grimly nicknamed 'Sherman neckties'.

The only Confederate force between Atlanta and Savannah was a single Cavalry unit and a few thousand Georgia militia, most of whom were in no way capable of standing up to the iron veterans of Sherman's army. They made one try at attacking them on November 22nd, and were swatted aside as effortlessly as a mosquito. That was it for the militia; they never even tried again while the cavalry made utterly ineffective pinpricks on Sherman's flanks.

Officially, Pierre Beauregard was in command in this area - the 'Department of the West', that rapidly shrinking area from Georgia to the Mississippi River - but because of Jefferson Davis' dislike and contempt for him, Beauregard's job now was primarily a logistical one, with no direct command in the field except in the most dire of emergencies.

The problem was that Beauregard's main field force was the Army of the Tennessee under John Bell Hood, and Hood - despite strong advice to the contrary - was determined to execute an invasion of eastern Tennessee that would at best be wasteful of precious men and supplies and at worst risking the only large formation left in that area.

In the event the worst case turned out to be far worse than anyone thought possible; the Army of the Tennessee was badly battered at Franklin on November 30th where Hood lost *fourteen* of his best generals and then finally broken beyond hope at Nashville on December 16th (something Beauregard didn't find out about until just before Sherman's forces slammed into Savannah), and Beauregard had no choice other than to take direct control of the forces in theatre that he could still contact and try to salvage what he could.

The helplessness of the people of Georgia was bad enough, and worse still was the utter helplessness of the Confederate Army, but worst of all were the 'bummers'. Mostly comprised of men who had semi-deserted their formations, they stole, burned, destroyed and in some cases killed as the Army rumbled through. Other bummers were Georgia unionists – mistreated by their neighbors for almost four years – and freed slaves who simply tagged along with the blue steamroller. The result in those cases was usually very rough justice indeed.[77]

On Thanksgiving Day, Sherman took the old state capital at Milledgeville, and rescued a handful of prisoners who had escaped from the notorious POW camp at Andersonville. These men looked like living scarecrows, and their appearance so angered the troops they burned the city to the ground after stripping it of everything they could carry away, and from that point on if anything the Federal forces were even more destructive.

Their goal, the port of Savannah, was now wide open despite the best efforts of the Cavalry to block and mine roads. That was enough for the CSA leadership in Savannah; they pulled the ten thousand troops there out and headed north. Sherman took the city on December 21st and sent President Lincoln a cheerful telegram:

---

[77] McPherson, pg. 810

The Long Patrol

"...I beg to present to you, as a Christmas gift, the city of Savannah, with 150 heavy guns and about 25,000 bales of cotton."

Sherman's troops entering Savannah, December 21, 1864. -- Sketched By Theodore R. Davis. Source: Harper's Weekly

Wilmington, NC, one of two big Atlantic ports left in Confederate hands, went down a month later in January 1865. That left Charleston alone, and the tolling of the Holy City's church bells only seemed to be counting down the moments until the end. Sherman moved out from Savannah on February 1st, and from there it was just a matter of time.

Sherman's engineers bridged streams and built roads through swamps the Confederates had declared impassable, and there was not a thing they could do to stop them. Sherman then pulled off a masterful feint – splitting his army and launching raids to make it appear that he was heading for both Charleston and the big supply center at Augusta, Georgia.

Beauregard was trying to keep Richmond informed of what was going on and by all accounts doing a good job, but in Richmond President Davis and his advisors were blinding themselves to the truth and for the most part refusing to believe that any army could be moving as fast as Sherman was without a supply line. Many of the senior members of the CSA as well as General Lee and President Davis themselves were West Point graduates, and they felt they had a good grasp of what could and could not be done by a modern army - but Sherman was not only working outside the book, he had completely thrown the book away as he inflicted Biblical levels of punishment on every inch of Georgia he marched through - "a pillar of fire by night, a pillar of smoke by day."

"...Far from stealthy." Fort Moultrie is evacuated, February 18th, 1865. (Photo courtesy the Library Of Congress)

It had to have been an awful moment when Beauregard realized that he would get no assistance or guidance from the leadership in Richmond, but he grimly assembled his troops for what he knew would be a last stand – and then realized too late that Sherman wasn't heading for either city, but was instead continuing straight up the center of the state for the capital at Columbia, which very nearly met the same fate as Atlanta on February 17th. When Beauregard found out what had actually happened, he had no options left. If he didn't get out *now,* he and his troops would be trapped quite literally between the Devil

The Long Patrol

and the deep blue sea. His troops got clear, but just in time. Beauregard's reward was to be relieved of his command by President Davis five days later.

On the afternoon of the 17th, while Columbia still burned and one year to the day since the *Hunley* had sailed out to take on the *Housatonic*, Union signalmen posted around Charleston intercepted a message sent from Beauregard's headquarters to Sullivan's Island: *"Burn all papers before you leave."* [78]

The next morning, Sullivan's Island was evacuated with anything but stealth, the huge magazine at Battery Bee being set off in an explosion felt across the harbor. Battery Marshall sat quiet and deserted, its guns spiked. A few miles to the south, the ironclad USS *Canonicus* decided to go for broke and enter the harbor. Not a single shot challenged them and they pressed on towards the symbol of Confederate control, the battered remains of Fort Sumter. A couple of broadsides went unanswered, and then a few minutes later a small boat carrying a white flag came bobbing out to meet the *Canonicus* and other monitors that were closing to join her.

The USS Canonicus. This picture, taken in 1907, shows she has survived in her Civil War configuration. She was finally stricken in 1908. Source: U.S. Navy Historical Center.

---

[78] Burton, pg. 317

CSS H.L. Hunley

The siege of Charleston ended that day, with the Stars and Stripes raised over Sumter for the first time since Pierre Beauregard's artillery had battered it down in April of 1861. Upwards of one hundred Federal ships entered Charleston Harbor before sunset, led by Admiral John Dahlgren. They found dozens of wrecked CSN vessels including *Chicora*, blown apart by her crew, and the *Indian Chief*, which may have been sunk by a Singer torpedo that got loose from its moorings.

One wonders what the Admiral thought as he passed *Housatonic's* wreck, already badly worm-eaten, or if he even thought about the black iron boat a thousand yards away. Probably not – victory can make one forget a great many things.

Time passed.

The *Hunley* passed into the realm of near myth over the years as her hull proceeded to fade further and further into the silt off Charleston Harbor. [79] Based on studies done since her discovery, she was probably mostly buried within ninety days, and within thirty years, there would have been no trace of her visible on the sea bottom – but that puts us a little ahead of the story.

By 1870, work was underway to clear Charleston harbor of not only the dozens of wrecks in the vicinity as well as the torpedo fields. It was at this time that the legend of the *Hunley* being dragged down by her intended victim really got started. Part of it came from the statements of Chief Engineer Tomb and Captain Gray, who strongly believed that being pulled into the torpedo hole by water suction was a realistic possibility, while some of it came from comments made by divers who had been down to the *Housatonic's* wreck – and said they'd seen the *Hunley* themselves.

The divers said in 1872 that they had not only found the *Hunley* but turned her propeller and saw the skeletons of the crew inside. That particular story, however, has the unpleasant

---

[79] The dozens of vessels that were sunk in and around the harbor in an effort to block it sank with surprising speed into the sea floor and were almost completely gone by the end of the war – and these were vessels several orders of magnitude larger than *Hunley* - MJK

whiff of manure about it. As Burton points out, the divers in their bulky suits and helmets could never have gotten *through* the narrow hatches, much less looked inside them.[80] We also know that *Hunley* would have been covered in silt by that point, and that there were no openings in the hull that could have made the interior visible – and even if there had been, the crew's remains would have been gone, not found at their posts in 2000.[81] Technically, however, the sank-beside-her-victim meme was at least possible, and had the added bonus of being morally comforting and uplifting.

The most authoritative comment on the subject, though, came from Will Alexander himself. Alexander said after the war (and later put in writing) that he felt the momentum of the collision with *Housatonic* dragged the *Hunley* into the now gaping hole in the cruiser's side and took her to the bottom. Alexander compounded the error by stating unequivocally that the boat had indeed been found by Federal divers beside the wreck. The reality, of course, was that the *Hunley* was nowhere even close to the hulk, otherwise she would have been spotted when Federal divers did their survey on February 20th. But for some reason this story continued to live on and grow for decades – I vividly remember reading it in submarine histories as late as the 1960s.

It should have been shut down by 1870, when the Federal Government decided that *Housatonic's* wreck was simply too close to the surface to allow for safe navigation. Divers did another survey, and still they found no submarine locked in a death grip with her prey. The Corps of Engineers then proceeded to blast, dredge, and salvage the cruiser's wreck almost out of existence, a process that continued in one form or another until 1909. But in all of that, no one ever found the *Hunley*.

---

[80] Burton, pg 238.

[81] In the spirit of fairness it should be pointed out that *Housatonic's* boilers – very roughly the same dimensions as the *Hunley* – would have been blown out of her fractured hull and onto the sea bottom. Divers who saw those boilers may have believed they were looking at the beast that sank her. Human nature would have done the rest - MJK

CSS H.L. Hunley

People never quite forgot about it, but as time passed the memory took on a life of its own. The divers who had twice pulled her out of the brown muck of Charleston Harbor claimed they had seen her, but never offered any proof. Mark Ragan found a reference to the one and only P.T. Barnum offering $100,000 for the *Hunley*,[82] but apparently the Master Showman miscalculated the number of suckers available that particular day; no one ever claimed the reward.

Time passed.

USS Holland (SS-1), the first practical combat submarine, in port during the early 1900s. Source: U.S. Navy

---

[82] Ragan, pg. 209

The Long Patrol

She lay there, quietly silting over as the world went on without her and her crew. William Alexander made a quiet name for himself keeping the memory of his friends alive and reminding the world that the *Hunley* had set the world's navies on a new path. The first true submarines – self-propelled with gasoline or diesel engines and capable of launching torpedoes, not ramming their targets with them, came into service. USS *Holland* (SS-1) joined the USN in 1900, and submarine technology took off from there.

By the time the First World War started in August 1914, all the major combatants had them – but only Imperial Germany had any idea how to use them. At first they were used as raiders, surfacing to mostly use light gunfire on their targets and then giving the crews a chance to get off in their boats – often even broadcasting the location of their victims to rescue ships. But eventually they stalked their targets from beneath the surface and fired on them without warning – and in the process very nearly starved Britain out of the war.

A generation later, Nazi Germany came even closer than the old Imperial Navy did, and this time the threat didn't end until the last U-boat raised the white flag in May of 1945.

The US Navy learned its lesson well, and its fleet boats sailed from Pearl Harbor and Australia to blockade and strangle the Japanese Empire – a job they did so well that by the time the Pacific War ended, the Japanese merchant marine had been reduced to a handful of coasters that dared sail only by night between the Home Islands, where the submarines couldn't reach. Out of twenty-three hundred ships in Japanese merchant service at the beginning of the war, only two hundred and thirty one survived nearly four years of assault by the Silent Service and most of those dared not even leave port for fear that an unseen raider would kill them.[83]

Within ten more years, the first nuclear boats were in service – sleek, silent warriors whose underwater endurance was limited

---

[83] And staying in port wasn't always a guarantee of safety; there are documented instances of US submarines torpedoing ships tied up at dockside and in one memorable instance 'torpedoing' a railroad bridge and destroying the train passing over it - MJK

CSS H.L. Hunley

only by the amount of food they could carry for their crew. An irascible Navy engineering officer named Hyman Rickover (whose nickname 'the Kindly Old Gentleman' belied his thoroughly dictatorial but highly successful style of management) almost single-handedly pushed the project through in the early Fifties with USS *Nautilus* (SSN-571), followed by USS *Seawolf*, the *Skate* and *Skipjack* classes, and then the modern, impressively lethal hunter-killers that still serve today. There is the *Los Angeles* class, backbone of the submarine fleet, a new *Seawolf* class – only three boats, but the three finest in the world – and the new *Virginia* class that will soon join the fleet. But even as the first SSNs began to come on line, interest in the *Hunley* started to rise again.

USS Virgina, SSN-774. The nuclear-powered hunter-killer is a far cry from the H.L. Hunley. Source: U.S. Navy.

The reason for that was the approaching centennial of the Civil War in 1961. For a brief time, everything Confederate or Federal was in vogue, and new generations of Americans learned of the *Hunley* and what she accomplished that cold February night off Charleston. Most still believed that she was dragged down by the sinking *Housatonic*, but by that time the old legend had been pretty much put down. Now the question was, if the *Hunley* wasn't alongside her victim, where was she?

The first dedicated expeditions to find her started a few years later, but most worked in the wrong direction and the wrong places. Various searches, both fully funded organizations and informal groups went at it on a stop-and-start basis for years. Edward Lee Spence of South Carolina led a search project claiming that he found the *Hunley* and actually dove on the wreck in 1970. However, for legal, technical, and scientific reasons (concisely laid out by Mark Ragan[84]) it appears unlikely that he actually did so.

For a while, the missing submarine was treated almost as more of a curiosity than the historical treasure it is – Jack Grimm, who led expeditions to find the *Titanic* (and nearly found her), offered a cash reward for her propeller. It is difficult to imagine why someone would have offered a reward for dismembering the *Hunley*, but there it is. After that – and the condemnation it brought – interest died down a bit until 1980. That year, the man who would ultimately find the *Hunley* decided to get into the game.

Dr. Clive Cussler – best-selling author, explorer, and buccaneer of the finest kind – had written several novels that involved a fictitious Federal agency called the National Underwater Maritime Agency (NUMA). When Cussler decided to start looking for famous shipwrecks on his own, he and a group of close collaborators set up a real-life NUMA and jumped in with both feet. On the other hand, Cussler would probably be the first one to tell you that what he ended up jumping into wasn't quite what he had in mind.

His first attempt was trying to find John Paul Jones' ship, the *Bonhomme Richard*. That and almost sixty subsequent surveys have turned into adventures worthy of a book on their own, but the *Hunley* seemed to have a certain attraction for Cussler and in 1980 he traveled to Charleston for the first time with a crew of like-minded brigands. They did find the *Housatonic* and surveyed her thoroughly, but no *Hunley*.

The following summer, Cussler tried again, and this time his team managed to find ships like the *Weehawken* and the *Keokuk*

---

[84] Ragan, pg. 234

in their shallow graves off Charleston, and the legendary blockade-runner *Stonewall Jackson* – which turned out to be resting under what is now a street on the Isle of Palms. There was one contact that for a few minutes had their hearts stopped – a smooth black metal cylinder whose dimensions appeared to match what was known about *Hunley* – and then turned out to be a Coast Guard buoy that had headed for the bottom long ago. Thus endeth the 1981 expedition.

Cussler would wait thirteen years before going after the *Hunley* again, while providing selfless and valuable assistance to Mark Ragan in his early 90s expedition. Ragan and his team found some promising contacts, but in the end the *Hunley* escaped again. In the meantime, Cussler and his team located other famed vessels. There was CSS *Florida*, the raider whose exploits so enraged the US Navy that they risked starting a war with Brazil to capture her, and then sank her in the James River. Cussler and his men found her, but the USN decided that any artifacts that had been recovered belonged to them and confiscated them. [85]

There was the CSS *Louisiana*, the unfinished ironclad that had gallantly tried to block David Farragut's assault on the forts below New Orleans. He found her, but like *Stonewall Jackson* she now lies beneath the shoreline in front of Fort St. Philip, one of the forts she tried so hard to protect. Cussler even found the remains of USS *Akron*, one of the US Navy's two giant aircraft-carrying dirigibles, on the sandy Atlantic floor off the coast of New Jersey. But something kept calling him back to the Palmetto shore, and in 1994 Cussler came back, this time determined to find the *Hunley*. The *Hunley*, however, seemed just as determined not to be found.

From the beginning, the '94 expedition seemed to have a

---

[85] Cussler, *The Sea Hunters*, pg. 98. The artifacts Dr. Cussler recovered are on display at the Nauticus Center in Norfolk, VA; with a small placard that does indeed identify who recovered them. However, the USN is notoriously unpleasant with *anyone* who recovers artifacts without their official imprimatur and as Dr. Cussler is not the sort of man who suffers fools and/or bureaucrats easily, the collision between the two was neither pretty nor pleasant. This and the story of how the *Hunley* was ultimately found are taken from Dr. Cussler's books and the NUMA website, www.numa.net – MJK.

The Long Patrol

McHale's Navy quality about it. The South Carolina Institute of Archaeology and Anthropology provided a dive boat and rustled up a platoon of sport divers whose well-intentioned eagerness far outweighed their skill and abilities. The SCIAA team leader seemed to believe that with the professionals on the case, the *Hunley* would reveal itself posthaste – so much so that he kept repeatedly identifying unknown targets as the lost boat, regardless of their size or dimensions. Marker buoys were repeatedly lost, and one of the divers almost died on the bottom before being rescued by one of Cussler's team.

They pressed on for a little bit longer, and then Cussler reluctantly gave up the hunt and went home to try and figure out what had gone wrong. After reviewing their clues over and over again, Cussler first came to the conclusion that LTC Dantzler's memo of February 19th couldn't be completely relied upon. He then took a good hard look at Seaman Fleming's testimony to the Board of Inquiry, and realized that *Hunley* had to have survived the attack on *Housatonic* and was pulled further out to sea by the ebb tide, which still had half an hour to run at 2130. Notifying his team of what he'd discovered, Cussler stayed home in Colorado while the NUMA divers made occasional sweeps further out to sea from the buried hulk that had once been a sleek cruiser. And on May 4th, 1995, Cussler got a very early-morning phone call.

Without question, *this* time the Blue Light had finally been answered.

She was there, intact and in one piece, at least as nearly as they could tell without a full excavation. The NUMA divers cleared away the mud around one hatch and discovered that one of the forward deadlights was missing, as well as a hole just below the forward hatch coaming. A quick examination revealed that the hull seemed to be packed solid with silt, which meant that anything inside the boat was preserved, and possibly very well so. The divers took pictures and videos that proved beyond any doubt that at last, the *H.L. Hunley* was about to finish her patrol. And with one last victorious flourish, they left a note tucked into the hatch: *"Veni, vidi, vici, dude!"*

CSS H.L. Hunley

Cussler had no intention of doing anything other than find the *Hunley* and turn the data over to the Proper Authorities. Unfortunately, the Proper Authorities got into a monumental fight over exactly who owned the boat and what was going to be done with it – and Cussler ended up smack in the middle. South Carolina said it was theirs, and a secondary fight erupted between the Holy City and the capital at Columbia over who'd get her.[86] Alabama, her birthplace, said that by rights she should be back at Mobile - perhaps alongside the battleship *Alabama*, within sight of where Park and Lyons had once stood.

Then the five-hundred-pound gorilla stepped in: the Federal Government, which had lost one of its prized warships to the *Hunley* one hundred and thirty-one years before, announced that since all former Confederate property came under the aegis of the General Services Administration the *Hunley* belonged to Uncle Sam, and the good curators at the Smithsonian and the US Navy Museum at Washington Navy Yard began to salivate over who would get possession of an artifact nearly as priceless as *Old Ironsides* herself.

In the meantime, Cussler and his team quite rightfully refused to release *Hunley*'s actual position to anyone until the mess was straightened out. SCIAA demanded that the wreck site be turned over to them immediately – and its head then stated that a buoy should be anchored over her. As a potential black market in *Hunley* parts had already arisen – though possibly only by rumor – that would have been a final death sentence for the boat. While two states, three cities, and the Federal Government argued over who would get her, others with few scruples and less respect for what the *Hunley* symbolized would strip her down to her component parts and scatter them to the four winds.

Cussler's refusal to reveal the wreck location until he knew she was safe was badly misinterpreted as an effort to extort money, ravage a Confederate war grave, keep the boat for himself, or all of the above. If the old saying "No good deed

---

[86] The resort city of Myrtle Beach jumped into the fray as well, but the idea of *Hunley* resting in honored glory among the fast food places, miniature golf courses, strip bars and theme parks did not gain much support - MJK

goes unpunished" ever needed an object example, then behold Clive Cussler and NUMA[87].

It took a while for things to finally get sorted out – along with a very firm push from South Carolina's Congressional delegation, led by the redoubtable Senator Strom Thurmond – but in the end the Federal Government was going to make the call, and their call was an interesting one. Strictly by the book, the *Hunley* became property of the United States Navy, a fact that must have had James McClintock and Horace Hunley spinning in their respective sepulchers. This is not as odd as it sounds – with only a couple of exceptions, the many retired warships that are on display around the US are still the property of the USN.[88] Then, South Carolina Senate President Pro Tem Glenn McConnell formed a commission to raise, restore and eventually display the *Hunley*, and the USN in turn turned the boat over to the *Hunley* Commission.

Now the real work started. First, *Hunley* had to be properly surveyed to determine whether or not she even could be raised – that call was far from a sure thing. That job fell to the divers, archaeologists and scientists of the National Park Service's Submerged Cultural Resources Unit. These extraordinarily talented men and women had been entrusted in the past with the demanding and dangerous survey of the shattered warships at Bikini Atoll in the Pacific, including the largest divable wreck in the world – that of the aircraft carrier USS *Saratoga* (CV-3), more than eight hundred feet of seventy-year old warship that was still carrying live explosives.[89] A less spectacular, but far

---

[87] In the end however USN did the right thing, albeit grudgingly: the Naval Historical Center officially acknowledges Cussler and NUMA as the finders of the *Hunley* - MJK

[88] For example, the battleships *Massachusetts*, *Iowa*, *New Jersey*, *Missouri*, *Wisconsin*, and *North Carolina* – as well as the *Texas*, which is nearly a century old – are still officially USN assets. In addition, some historic wrecked aircraft found in recent years have been allowed to deteriorate further because the USN claims they are still Navy property and cannot be recovered without their permission – MJK.

[89] *Saratoga* still has hundreds of live bombs and rockets aboard, some fully fuzed and armed, placed there to see how well they would survive on a warship under nuclear attack. Properly surveying the wreck required divers to get in close to and in the midst of that deadly cargo – MJK.

CSS H.L. Hunley

more sensitive survey was that of the battleship *USS Arizona* (BB-39) at Pearl Harbor.

In the 1980s, concerns about the physical condition of the wreck – a war grave with more than eleven hundred sailors still resting inside her – made it imperative that the *Arizona* be thoroughly surveyed. The SCR divers pulled it off magnificently, and provided invaluable information needed to preserve the eighty-year old wreck from becoming an ecological time bomb due to oil still trapped in her deteriorating hull. They were the perfect choice to do the survey on *Hunley*, and they did not disappoint.

Hunley was sitting about thirty feet down, under roughly a foot or so of silt. She was intact to an extent no one had even dared believe possible, save for the broken deadlight and the hole in the hatch. She was covered in 'concretion' – a hard shell of rust, seawater deposits, and marine life – that was attached to her like a second skin. It had preserved her beautifully, but there was a catch: any crack in the concretion would permit almost immediate deterioration of the boat's hull. It was determined that *Hunley's* condition had stabilized to the point that unless the recovery and restoration team was very, very careful, the boat would suffer from more corrosion in just six months than in all the years she'd been on the bottom.[90]

While *Hunley* was being mapped and surveyed, an incredible facility – very much a real, genuine time machine – was being assembled a few miles away. The old Charleston Navy Yard is full of early 20th century buildings with the incongruous sight of massive cruise ships docked across the road, but at the far end is a much more modern building – built and never used because in its infinite wisdom, Congress closed the Yard before it was ever put to work. But when a combination storage and restoration facility was needed for the *Hunley*'s trip back from 1864, it was ready and waiting. Eventually christened the Warren Lasch Conservation Lab (after one of the leaders of the effort to bring the *Hunley* home), it is beyond question the most advanced and capable lab of its kind in the world, and even after the *Hunley*

---

[90] 'Recovery', *Friends Of The Hunley* (www.hunley.org)

leaves it will remain a world center for its work.

A huge fresh-water tank was built to take her once she was raised, and the tank has been instrumented and monitored to an extent that would make a space shuttle blush. The tank is surrounded by a platform that allows easy access to anything in it (perhaps a bit too easy; not long after public viewing was permitted someone dropped an unauthorized camera into the tank) and allows constant monitoring of whatever is being restored. State of the art computers are studded throughout the building, some of which are connected to a massive x-ray machine that can scan anything from *Hunley*-sized objects to human remains.

And tucked away in one corner were nine refrigerated morgue drawers. They were there for Lieutenant Dixon and his men, but future artifacts brought to Lasch may also have remains in them, and it would be prudent to be prepared. [91] It took time to put the Lasch facility together, but that was something they had – *Hunley* wouldn't be going anywhere anytime soon.

It would take five years from discovery to salvage to sort out who owned the boat, what would be done with it, how to get it out of the water, and more importantly who'd pay for raising it.

That turned out to be the easy part – getting the *Hunley* to surface again was going to be something else entirely. They looked at a cofferdam to be built all the way around the boat and then excavate her – a good idea and a practical one, but hideously expensive. They considered simply putting slings around the sub and winching her up, but there was no guarantee that her rivets would hold together. The idea of simply dropping a giant clamshell around her and bringing the whole thing up was even looked at, but the seabed would have likely given out, possibly crushing the *Hunley* into a hundred and thirty-two year old soda can.

In the meantime, the boat – now under 24/7 watch – was being slowly and steadily excavated. More damage was found, gashes in both the bow and stern in addition to the damage

---

[91] 'Conservation', *Friends Of The Hunley* (www.hunley.org)

already found to the hatch. At the last minute, it is discovered that the torpedo spar is still bolted to the base of her prow, and there is a momentary panic as archaeologists try to figure out how to remove it. In the end, a simple wrench is used to remove it, the same way one was used to install it long ago. The damage to the hull complicated matters, for now the structural integrity of the hull was in serious question.[92]

What eventually saved the day was the discovery of a chemical foam that would be injected into a series of bags slipped beneath the *Hunley* and then gently push it upwards. Once this was done, *Hunley* would be strapped to a steel truss that was dropped precisely over the boat. The truss – now securely grasping the boat along its entire length – would then be pulled upwards and placed on a barge, then towed to the Lasch lab and transferred to the freshwater tank. It is an elegant, affordable, and practical solution.

All it has to do now is work.

August 8th, 2000, 0830 local. The *Hunley* has been on patrol for approximately one hundred and forty six years, five months, twenty-one days, thirteen hours, and thirty minutes. A flotilla of pleasure boats, tour boats, press boats, and just about anything else that will float has gathered around an odd-looking ship called the *Karlissa-B*, now moored just a few feet away from *Hunley*. The three-hundred ton capacity Manitowoc crane that is firmly anchored to *Karlissa's* deck is more than capable of lifting the truss and the boat cradled inside her, and it's just as well – nobody knows exactly how much *Hunley* weighs this morning, and overkill is probably a very good idea.

The lift crew is keeping a close eye on the weather – the lift

---

[92] This damage – including the forward hatch damage – was almost certainly incurred as salvage, search and wreck clearing operations went on around the *Hunley* from 1865 until the early 20th century. Had she suffered any of the damage during the attack, she never could have survived as long afterwards as she did. Salvors and engineers frequently used plows, chains, and even dynamite simply thrown over the side. *Hunley* lay in the midst of an area littered with wrecks and rubbish, and as no one realized she was there – remember, most people believed she was much closer to or even beside *Housatonic* – she was probably battered unknowingly over and over again through the years. - MJK

The Long Patrol

was supposed to begin at 0800, but gusty winds have flared up out of nowhere, as if in one last attempt by the Gods to keep the *Hunley* in her grave. But by 0830 even the Gods have decided that enough is enough, and it's time for her to come home. The lift supervisor gives the signal, and the heavy steel cables running down into the water writhe, quiver, then stiffen and start to move. Divers on the bottom back off as the truss suddenly moves, with a cloud of sediment suddenly erupting up around the *Hunley*. She moves, hesitantly at first, as if startled from a long sleep, then settles down and moves towards the light.

The Long Patrol, nearly over: *CSS H.L. Hunley surfaces for the first time in one hundred and thirty six years. (Photo courtesy of the United States Naval Historical Center, Washington, DC.)*

The *Hunley* was the first submarine to meet her end in the black, unyielding depths, but others followed over the years - boats with names like *Thresher* and *Scorpion*, *Affray* and *Truculent*, *K-219* and *Kursk*.

Their crews died at their posts, slowly from inexorable suffocation or by lightning-fast collapsing bulkheads and fire.

CSS H.L. Hunley

Like the *Hunley's* crew they all knew what was coming, they all had time for one last thought of home and loved ones, a final prayer, or a terrified scream. All of them though would have wanted to hear one final command, the one *Hunley's* crew will now hear after one hundred and thirty-six years:

Surface.

At 0839, the truss moves steadily upwards through the waves and into the air. The sounds that echo across the harbor are incredible – whistles, ships' sirens, cannons, and cheers – my God, the *cheers*!! - that seem to roll over the waters. If the shades of George Dixon and his men ever had any doubt as to how they have been regarded by history, they cannot possibly have any now.

Hunley sways gently for a few moments as the lift team gets the truss stabilized, and then she is carefully swung over a barge for the final trip upriver to the Charleston Navy Yard. The cheering is still going on, a roar that will not end until the boat is safely ashore.

It takes just a few minutes to secure the truss and its cargo, then the tug that will tow the barge back upriver gives two quick toots on its whistle as she begins to move. If anything, the cheers grow louder as the onlookers realize now that the *Hunley* is, indeed, coming home and nothing can stop it.

Clive Cussler, beaming from ear to ear, turns to reporters on the press boat *Carolina Clipper* and says, "Well, I have to be going now", then dives over the side and backstrokes out to the nearby *Diversity*, the boat that NUMA was using when they found *Hunley*.

The barge, followed by dozens of small boats, proceeds up the harbor at a calm, dignified pace. They pass the carrier *Yorktown*, a ship whose use and sheer size would have been dismissed by McClintock and Hunley as the wildest of fantasies. This morning though, *Yorktown's* flight deck is not crowded with aircraft but instead with thousands of cheering onlookers, applauding and waving Confederate flags as the *Hunley* cruises past.

The Long Patrol

Eleven of those onlookers are remaining silent – nine women in widows' weeds, one for each of the men still resting inside the boat, and two men at attention in Confederate uniforms, flanking a lantern with a blue light.

VII: VALEDICTORY

Battery Marshall was abandoned at war's end and eventually demolished. The Battery ran more or less along the eastern shore of what is now Marshall Boulevard on Sullivan's Island. A historical marker stands at the south end of the Palm Boulevard bridge (see below), and a visitor can walk the beach at sunset and see the view George Dixon had on February 17th. The remains of a dock can still barely be seen at low tide, and it's not at all hard to close your eyes and see the black iron monster sitting there as her crew quite literally prepared to do or die.

General Pierre Gustave Toutant Beauregard, CSA, went home to New Orleans and a hero's welcome after the war, but it was a close-run thing. He did not take his relief after the evacuation of Charleston well and he ended up subordinate to GEN Joseph Johnston, himself trying to overcome disgrace after playing a key role in the loss of Atlanta and who was now given the awful task of trying to hold together what was left to the south of Robert Lee's collapsing front. In theory this was made up of three field armies including the battered Army of The Tennessee, but in reality barely enough men were still fighting to equal two solid divisions. Johnston came up with a plan to join forces with Lee and then strike back at both Sherman's spearheads and Grant's Army Of The Potomac.

Beauregard had little if anything to do with the planning; he was handling routine staff work and keeping it off Johnston's back. In the end it didn't matter - in less than sixty days, it was all over. Johnston and Beauregard met with Sherman on April 17th, 1865, near Durham, NC, and after three days of talks they surrendered all Confederate field forces in the Carolinas, Georgia, and Florida - just a little shy of ninety thousand men. Lee's surrender eight days earlier was the one everyone remembers, but Johnston's was the one that for all practical purposes ended the war.

The Long Patrol

For that honor, President Davis would accuse both Beauregard and Johnston of treason. Beauregard himself surrendered to Sherman on April 26th, and a few days later headed home to New Orleans.

Beauregard pursued several careers afterwards - he considered an unusual offer to take command of Brazil's armed forces, but that didn't work out. He went into business and did well as an administrator, but was run out of successful jobs twice by business matters he had little or no control over.

But in 1877 he became a manager of the Louisiana Lottery, along with his former comrade Jubal Early. He was paid well - his image and reputation were a great selling point -and invested better, making him a wealthy man. Beauregard did a great deal of writing on history and the recent unpleasantness, and carried on a bitter feud with Jefferson Davis that ended only with Davis' death in 1889. Beauregard was asked to lead the former President's funeral procession but declined, saying that "I cannot pretend I am sorry he is gone."[93] Beauregard himself held on for another four years, dying of heart disease in 1893. He rests in the Army Of The Tennessee's Louisiana Division tomb in the Big Easy's legendary Metairie Cemetery - a dignified green mound topped by a statue of General Albert Sidney Johnston, killed so early in the war at Shiloh. One would hope that no one ever remembered the fact that Pierre Beauregard so disliked being under any other general.

Dr. Clive Cussler is still writing books, still searching for lost ships, and still having the time of his life. Long life and good fortune to him.

Just before 0600 on the morning of August 5th, 1864, Admiral David Farragut, lashed to the rigging of his tough old flagship *Hartford*, ordered the United States Navy into *Mobile Bay*. The two forts that guarded the main approach to the harbor opened fire with everything they had, and the *Hartford* staggered

---

[93] Hattaway, Herman M., and Michael J. C. Taylor. "Pierre Gustave Toutant Beauregard." In Leaders of the American Civil War: A Biographical and Historiographical Dictionary, edited by Charles F. Ritter and Jon L. Wakelyn. Westport, CT: Greenwood Press, 1998.

CSS H.L. Hunley

repeatedly, but led the other thirteen ships in the task force through. Now, however, they faced another challenge – Mister Singer's torpedoes. One of them tore a hole in the monitor USS *Tecumseh*, sending it almost straight down to the shallow floor of the bay. The other ships – except for *Hartford* – hesitated for a moment, until Farragut got a look at them and bellowed, "Damn the torpedoes, full steam ahead!!" The Admiral was not disappointed; the task force formed back up and pressed on, pausing for breakfast once they were safely in the bay.

It was still a long way to Mobile itself, and CSA Admiral Franklin Buchanan, still in command of Mobile's naval defenses, had no intention of giving in without a fight. He ordered out what was literally his secret weapon: the ironclad CSS *Tennessee*. At the time, the Confederate ironclad was probably the toughest warship afloat – but its strength was bought at the price of almost no speed or maneuverability. Farragut watched *Tennessee* move out from their anchorage at the head of the bay and head directly for the fleet. The other officers watched in trepidation, but Farragut merely smiled and commented that he hadn't thought old Buck was such a fool.[94]

Taking personal command of the *Hartford*, Farragut led the USS *Lackawanna* and one other ship to greet the *Tennessee*. When they got into position, Farragut and his ships started ramming the *Tennessee* every five minutes. Unable to steer, unable to run away, and unable to maneuver quickly enough to get a good shot at its tormentors, *Tennessee* endured ninety minutes of this before she was reduced to a drifting hulk, and the few surviving CSN vessels retreated back upriver to Mobile itself. By August 26th, combined USN/US Army operations had taken the forts and secured the entrance to the bay once and for all. The last open Gulf port east of Texas was now gone, though the city itself would remain unconquered until the end of the war.

General Dabney Maury was one of the last CSA commanders to give in, not actually surrendering until after Lee's surrender at Appomattox. Maury eventually became a

---

[94] Catton, pg. 195

diplomat, becoming Ambassador to Colombia for four years. He died in Illinois in 1900.

Breach Inlet no longer flows unimpeded to the sea. Some years ago, South Carolina built a two-lane bridge that now carries Palm Boulevard across from Sullivan's Island to the Isle Of Palms. Quite a few famous names cross that bridge on a regular basis – the Isle of Palms is now a *very* high-end resort island for the rich, famous, and powerful. NASCAR drivers, US senators, and retired generals now call it home.

Fort Johnson was nearly a century old when the Civil War started, and four years of bombardment did it no favors. The fort itself has long since vanished, with an NOAA station currently on the site.

Fort Moultrie was rebuilt in 1870 - battered as it was, it was still in far better condition than the rubbish heap that had been Fort Sumter. Eventually the fort was upgraded to the point where it became the primary coast defense post for Charleston. It continued to serve in that capacity until 1947, when it was decommissioned and became a National Historical Landmark. Since largely restored to its Civil War appearance, Moultrie is a fascinating time capsule of three different styles of coast defense technology and weapons, and the WWII era Harbor Command Post is a stunning walk back in time. But walk back out to the Cove, and squint a little bit…there, among the fishing boats and pleasure craft James McClintock is getting the *Porpoise* ready for sea, with Charleston and Mount Pleasant stretched out behind him.

Fort Sumter remained an abandoned pile of shattered brick until 1898, when someone thought it would make an excellent site for a coast defense battery. The fort, originally three tiers high, was cut down to two then filled in with a reinforced concrete wall taking the place of the areas that couldn't be salvaged. Two 12" rifles were installed and christened Battery Huger, after a Revolutionary War general from South Carolina. Battery Huger survived until 1943, when it was decommissioned, but several 90mm guns for anti-aircraft and anti-torpedo boat (!) use remained in use until the end of the

Second World War. Decommissioned for the last time in 1948, it too is now a national monument. Restoration work continues, and today thousands of tourists take the boats from Charleston and Patriot's Point out to see the place where the War started.

James McClintock continued to work on torpedoes for the Confederacy until the end of the war. He developed – and kept – a thorough and complete hatred of the Federal Government, one not at all helped by the loss of his friends in Charleston. He was at Mobile when the war ended, and within a year or two was back to his old job of ship captain. In 1872 however, he made a very quiet – and secret – trip to Canada. McClintock was on his way to meet with a group of officers from the Royal Navy, and his intent was to give them all the information he could on the most successful submarine ever built up to that time – the *Hunley*.

By the book, McClintock could have faced some very serious charges if he'd been caught (Rich Wills points out that if he'd signed a loyalty declaration in 1865 - which he probably did – he could have been looking at treason charges), so he apparently had a great deal of motivation for going. He went aboard HMS *Royal Alfred* in Halifax Harbor, and spent almost three days giving officers of Her Majesty's Navy everything he had – and making it very clear that he had no compunctions about the RN using this technology against the United States.

Two British researchers discovered the official reports and intelligence summaries in the 1990s, and they are stunning. By the time McClintock was finished, he had given them everything they needed to not only build their own *Hunley*, but the information to correct all of its shortcomings. The RN officers were very impressed with what McClintock gave them, and said so in their report. But in the end, McClintock went home, the *Royal Alfred* returned to England…and the reports went into a dusty file cabinet somewhere in London, utterly forgotten.

One wonders what would have happened had the Royal Navy gone to work on a larger and more powerful *Hunley* with all the resources and science at their disposal, but it didn't happen – and forty-two years later, the capabilities of the

German *Unterseebooten* came as a grim shock. In any event, McClintock continued to work on diving and torpedoes until his unfortunate death in Boston Harbor in mid-1879 – while demonstrating a torpedo for the United States Navy. If nothing else, it proved that McClintock was flexible enough to work with the Enemy - if it was necessary to pay the bills.

USS Canandaigua finally came off the blockade line after Charleston's surrender. She was decommissioned in April 1865, but brought back into service that November. She served on the European station until 1869, and then came back to the New York Navy Yard for three years. For some reason her name was changed to *Detroit* for three months and then back to *Canandaigua*. She wouldn't sail again until 1872, and would spend the next three years quietly showing the flag in the Gulf of Mexico and the West Indies - somehow avoiding the grim fate legend says awaits as ship whose name has been changed. She was decommissioned for the last time on November 8th, 1875, and was broken up in 1884.

Admiral John Dahlgren, USN, ended the war a hero, but the rigors of four years on the line and the loss of his son took the spirit out of him. His courage and skill could never be questioned, and his sailors idolized him but his officers sometimes failed to live up to the standard he set. He would go on to command the South Pacific Squadron, the Bureau of Ordnance, and the Washington Navy Yard before his death in 1870.

The earthly remains of **USS Housatonic** still rest about three miles off Breach Inlet at 32° 43 08.75" N, 79° 46' 34.74" W. Nearly forty years of continuous dredging and salvage work left very little of the ship that was once the pride of the South Atlantic Blockading Squadron. When Clive Cussler got to her in 1981, all that really remained was her keel and lower hull, and what little was left of that is some distance beneath the mud and sand of the ocean floor.

In the amazement, celebration, and remembrance of the *Hunley's* resurrection it seems to me that *Housatonic* has been somewhat forgotten, and that is truly a shame. She was a tough, powerful ship with a solid, veteran Captain and crew who just happened to be in the wrong place at the wrong time one night and found themselves on the receiving end of history.

Perhaps once things have settled down a bit, some funds could be appropriated to do a full-dress survey of her resting place, with an eye towards bringing up at least a portion of what remains – hopefully as close to the torpedo impact point as possible. To see that displayed alongside *Hunley* would be to fully appreciate the magnitude of what Lieutenant Dixon pulled off that night, and the challenge to survive the unknown that Captain Pickering and his men endured. They deserve the remembrance as much as Lieutenant Dixon and his men do.

The submarine torpedo boat **H.L. Hunley** still rests in her tank of cool, fresh water at the Lasch Facility. Every day, devoted archaeologists and historians reverently pore over her, intending to uncover every one of her secrets. One hundred and thirty-six years of marine encrustation will be painstakingly removed. It has not yet been determined whether or not the gashes in her hull will be repaired, though she will be restored as closely as possible to operational condition. She will stay in that tank for a short while longer, then will move to a new restoration facility, most likely at Clemson University.

There is still some debate as to who found her – Clive Cussler or Edward Lee Spence. Both claimants have their partisans; both arguments have their merits. However, this writer comes down firmly on the side of Dr. Cussler. No offense or ill will is imparted to Mr. Spence, far from it. I believe he made his claim with the most honest of intentions and the utmost integrity, and I believe he believes that he found *Hunley*. But at the end of the day, it was Cussler who made the leap of reasoning to look in the right place and found her. *Q.E.D.*

And as befitting a craft that has left so many questions unanswered, the *Hunley* has asked us one more: in July of 2006, it was revealed that beyond doubt, the forward hatch was open, perhaps as much as half an inch. Did it contribute to her sinking? Unlikely at best, for reasons we've already covered. Did it happen after – or, more disturbingly, during the sinking?

Did the crew make one last desperate effort to get out? Did a stray grappling hook, scouring the harbor bottom after the war, snag the hatch and yank it open?

As this is written, scientists at Clemson are trying to determine the best way to actually preserve her. Several different methods are being looked at, and one leading possibility is something called supercritical fluid technology – a method where intense heat and pressure will be used to remove the lethally corrosive salts from her cast iron hide. Some rivets have been treated this way and although the jury is still out, the results look promising enough to warrant further investigation.[95]

She would remain at Clemson until around 2010, when she was to have made her last voyage – a trip to the specially constructed display building that will be her final berth. The magnificent Museum of Charleston made a strong proposal to display her there, and the Patriot's Point Naval Museum in Mount Pleasant also made a stand. Both locations would have been fine permanent homes for the world's first combat submarine, and the decision would have been a tough one. But as Charleston and Mount Pleasant scaled back their offers after costs started to mount, a dark horse emerged with a definite edge – the city of North Charleston, where she already lies and as of this writing her final home will be there.[96]

Wherever her last drydock, by the beginning of the next decade the technology available to display and explain the *Hunley* and her crews will be incredible – sound, light, and interactive displays that will get visitors as close as humanly possible to actually being inside the little iron boat. It will be an unforgettable experience to anyone who sees it. But the "*Hunley* Experience" – and surely that is what it will be christened – should not be allowed to overshadow what the tiny iron boat stands for. There is, however, a possible question mark overshadowing the *Hunley's* future.

---

[95] *Fluids could help preserve H.L. Hunley*, The Associated Press, November 14, 2004

[96] As of this writing (September, 2012) the Hunley is still safely nestled in her cradle at the Lasch facility. As investigating and excavating her is still underway, there is no plan to move her in the foreseeable future - MJK

CSS H.L. Hunley

In late 2005, the Friends of the Hunley were taken to court by the state of South Carolina over charges that the organization was violating the state Freedom of Information act. It's not entirely clear why the matter got this far, but Palmetto State politics have never been thoroughly coherent anyways and the Friends' counsel admitted in front of the South Carolina Supreme Court that they were going to have to open their records eventually. What effect this will have on the time schedule of the restoration and display is unknown.

More political shenanigans regarding *Hunley* became public in May of 2006, when *The State*, South Carolina's largest newspaper, ran a series of articles that documented almost non-existent oversight of the funds appropriated to raise and preserve her[97]. The original estimates of those costs done in 1995 came to approximately five to ten million dollars. Even given the – shall we say – flexibility of government estimates, the current (May 2006) estimate of $97,000,000 represents a nine hundred percent increase. It includes more than forty million dollars for the *Hunley* museum itself, more than thirty million for the facilities at Clemson, nearly twenty million more already spent or earmarked for her raising and preservation, and three million for Civil War memorabilia that will be displayed with her.

It is not impossible that the questions being asked about *Hunley's* funding could seriously cut back – or cut off – that funding altogether. Were that to happen, the only other organization that could take over and finish the job (for no matter what happened politically, it would be unthinkable to allow *Hunley* to rot into a pile of rusty iron plates and rods) would be the Smithsonian. That possibility would open a new war, this time one of words over who would get final custody of the black iron boat.

On top of those problems, the involvement of Clemson University seems to have been for less than altruistic reasons. Admittedly, no university worthy of the name lives purely for academics any more, and business comes first and foremost. But in this case, e-mails uncovered by *The State* seem to show that

---

[97] *The State*, May 14, 2006

money was running neck-and-neck with an overwhelming desire to be prominently mentioned in a projected Discovery or History Channel special on the *Hunley*, the idea being that said appearance would suddenly vault Clemson into the rarefied ranks of the best schools in the nation. (Doubtless there was also a strong urge to get one over on downstate rival University of South Carolina, which was prominent in many searches for and recovery of the *Hunley*.)

Regardless though of where one stands – and I assure you, even a hundred and fifty-one years later, you stand *somewhere* on the Civil War or its causes – look past the flags and the words and the regrets and the what-ifs. Look instead at Americans who found the skill and the knowledge and the resources to do something no one had ever done before, and even though they lost their lives in the process they pulled it off.

The boats that followed them, however, are their truest and longest-lived legacy. The *Hunley* pointed the way towards the *Holland*, the *Plunger*, and the increasingly sophisticated, capable and far more lethal boats that followed. The old Imperial German navy's first U-boats came terrifyingly close to starving Great Britain out of the First World War, and their sons in *die Kriegsmarine* got within an ace of doing it in the Second. Take a look at a map of merchant sinkings in the Atlantic – some of them were so close to Charleston that the *Hunley's* crew would have heard the hollow thud of torpedo warheads and wondered at the metallic crunch of collapsing bulkheads.

In the Pacific, the USN's submarine fleet – the only group of warships capable of hitting back against the victorious Japanese in the early days of the war – overcame serious technical and doctrinal problems to slowly strangle the Japanese merchant fleet. Their victory was so overwhelming that one can make a good argument that they could have starved the Empire into submission without the use of the atomic bomb – but that is a discussion for another place and another time.

Postwar, the Soviets, whose method of making war was once compared to producing grand opera, planned to flood the Atlantic Ocean with more than two hundred conventional and

CSS H.L. Hunley

nuclear attack boats to savage the convoys that would sail to save Europe a third time – and there would be nowhere near the numbers of ships or aircraft this time to defend them that there were in the Second World War.

This time, each ship would carry four to six TIMES the cargo of the old *Liberty* or *Victory* class – but each sub was at least that many times more dangerous than their German predecessors. A third Battle of the Atlantic would have been a short, murderous affair that would have differed from its predecessors in a very dramatic way – for the first time, other submarines would have been a major weapon against the attackers. Up to that point, the *Hunley* had set a century long precedent of using submarines against almost exclusively surface vessels (there were a handful of sub-vs.-sub battles but they could be counted on the fingers of both hands, more accidental encounters than anything else), but from the early 1960s on, things would be different.

The first SSKs - 'hunter-killer' boats capable of remarkable silence compared to previous subs - showed up in the late 1940s in the world's major navies. They almost invariably used improved technology intended for the last generation of U-boats, subs that might have tipped the balance in the Battle of the Atlantic had they been deployed in sufficient numbers. In the US, these boats later developed into the first nuclear sub classes and then the true hunters - the *Permits*, *Sturgeons*, and *Los Angeles* classes. In the Soviet Union, a bestiary of more than fifteen separate classes with a bewildering series of NATO code names ended up going to sea.

The next major departure was the development of the ballistic missile submarine, able to fire accurate (more or less, at first) missiles from any deepwater position within range. The early US boomers carried short-range, solid-fueled *Polaris* missiles whose accuracy and reliability wasn't good even in an era of questionable missiles, but were still enough to put Soviet planners into a cold sweat. The Soviet reply – which actually beat the first US boomer, *George Washington*, to sea – was to put a handful of even less accurate and more dangerous liquid-fueled missiles to sea, a solution that in the end killed at least one

The Long Patrol

Soviet boomer and may have maimed more, not to mention sending a few score Soviet submariners to deep, cold graves.

In the end, both sides went through a progression of boats and missiles – the US with *Poseidon*, and then the awesomely long-ranged and accurate *Tridents*. The Soviets countered with a wild array of missiles and boats that culminated in the massive *Typhoon* class – true undersea monsters capable of salvoing hundreds of warheads against the United States from 'bastions' protected by dozens of Soviet attack submarines.

The next great change came with the development of cruise missiles in the late 1970s and early 80s. The superb US *Tomahawk* – fast, capable of flying just a few feet off the ground, and hyperaccurate – turned *every* submarine into a boomer, and the possibilities so badly scared the Soviets that they insisted the *Tomahawk* be regulated by the last Strategic Arms Limitation Treaty (SALT).

The US submarine fleet is now down to about fifty boats of all types – a mix of *Los Angeles* and *Seawolf* class attack subs, along with a dozen or so *Trident*s, four of which will be turned into *Florida* class cruise missile boats – each one able to launch fifty or more *Tomahawks* at once. A new attack boat, the *Virginia* class, is just now entering service, and they will become the backbone of the force into the early decades of the twenty-first century.

As of this writing, the USN is beginning development of a new generation of missile boats, but they are not likely to enter service until at least the 2020s - if at all. Fortunately, the *Tridents* are still in superb shape and preternaturally quiet - officially, the USN says that none have ever been tracked by an adversary. The Russians have deployed a new boomer, the *Borei,* or *Yuri Dolgorukiy* class, for which they are making the usual extravagant claims.

Without question, they are an advance on the *Delta IVs* that make up the Russian boomer fleet (only a single *Typhoon* still serves), but they are essentially warmed-over late Cold War technology, up against continually advancing US and British technology. Add to that questionable budgeting, a balky missile, and the traditionally sketchy Russian quality and upkeep, and it's not likely that many will see active service.

No matter what they are called, however, they will continue their dark, quiet patrols, gliding from their lairs out into the deeps where they vanish unseen and unheard until they strike. In that, for all the technology and history that has passed since a cold February night in 1864, they are no different from the little iron boat and its brave crew.

The *First True Combat Submarine* came from an unexpected source - not a Government dockyard, but the pen and drafting table of an Irish immigrant who rather disliked the British.

The first truly practical modern submarine was designed by one John P. Holland of Paterson, New Jersey by way of County Clare, Ireland. Holland had emigrated here in 1873, and started designing submarines almost immediately. He submitted the plans to the USN, but they almost immediately rejected them as impractical. This would have been a setback to almost anyone else, but not to Holland - he had financial backing from the Fenian Brotherhood, an Irish nationalist group that felt Mr. Holland's designs might be just the thing to use against the Royal Navy.

Holland apparently had no moral qualms whatsoever about that, and eventually ended up designing a boat slightly smaller than the *Hunley* called the *Fenian Ram,* powered by an internal combustion engine and armed with a centerline-mounted pneumatic gun. It had no ballast system to speak of, submerging more through brute force than elegant engineering. It worked, after a fashion, as did a smaller version that Holland was using to work out control and stability issues. However, the issue of payment reared its ugly head, much as in the case of the *Intelligent Whale* - to be precise, the Fenians apparently had no intention of actually paying Holland for his work or his boats.

Discussions followed, apparently becoming more acrimonious until the Fenians terminated the negotiations by simply stealing the *Ram* and the smaller boat. A good idea, perhaps, but one dashed by the ugly reality that in fact, no one in

The Long Patrol

the Fenians was actually able to operate the boat. (The *Ram*, by the way, survived all of this and today rests in a New Jersey museum.) Holland understandably refused to assist them, and went on to refine his designs one more time into a fifty-three foot long boat he had built in Elizabeth, NJ.

For the first time, all the features of what we would call a modern submarine were in one place - an internal combustion engine for surface operation, an electric motor for submerged running, a well planned, reliable ballast system, and a true torpedo tube. But most of all - after the US Navy tested it to within an inch of its life for nearly four demanding years - it was safe enough that when it went out, its crew stood a very good chance of coming back. Modestly christened USS *Holland*, she was designated Submarine (SS) 1. She served well and honorably from 1900 to 1905, when she was decommissioned. *Holland* survived until she was sold for scrap in 1913, but by then she had been supplemented and replaced by progressively larger, faster, and more capable boats - not just in the USN, but in other navies as well.

The Royal Navy, in fact, bought five Holland-designed and built boats as their first submarines - proving that Holland, like James McClintock, was more than understanding when it came to dealing with the enemy. The USN's Holland boats, by the way, were built at Holland's own yards - the Electric Boat Company of Groton, Connecticut, today the Electric Boat Division of General Dynamics.

The major navies of the world, however, weren't at all sure what to do with the new invention. Most planned to use it as a scout, or for harbor defense. But Imperial Germany - looking for ways to challenge Britain's domination of the high seas and strangle her lifelines in the war they knew would come - had a different idea.

The ***Next Ship Sunk By A Submarine*** was the Royal Navy cruiser HMS *Pathfinder,* torpedoed by the German submarine *U-21* on September 5th, 1914 – fifty years and seven months since *Hunley* became the first to pull it off. *U-21's* skipper, Korvettenkapitan Otto Hersing, had to get in close too so as to be sure – in his case, under a half mile. His torpedo penetrated *Pathfinder's* steel hull just aft of her bridge and set off the forward magazine.

The resulting explosion shook *U-21* so hard that Hersing thought he was going to lose her, but after getting things back under control he went to periscope depth and saw the shattered remains of *Pathfinder* sliding under the North Sea. There were only eleven survivors out of a crew of over three hundred.

The Next Ship To Be Sunk By A Submarine: *HMS Pathfinder, torpedoed within sight of shore by U-21 on September 5th, 1914. There were only eighteen survivors from a crew of two hundred and fifty eight. (Photo courtesy Royal Navy)*

The world wouldn't have to wait fifty years for the next kill. On September 22nd, *U-9* under Kapitanleutnant Otto Weddigen put the fear of God into the Royal Navy by killing the cruisers HMS *Cressy*, HMS *Aboukir*, and HMS *Hogue* in the space of an hour. Less than a month later, Weddigen took out HMS *Hawke* while the cruiser was stopped to transfer mail. From there, the long list of ships just marched on and on. The submarine was a weapon now and it would only grow more dangerous.

The ***First Ship Sunk By A United States Navy Submarine*** was a nondescript Japanese freighter named the *Atsutasan Maru* off the Chinese coast on December 16th, 1941. Her executioner

The Long Patrol

was USS *Swordfish* (SS-196), operating from her doomed base in the Philippines. *Swordfish* and her crew had little time to enjoy the victory - eleven days later, she was evacuating the Submarine Force staff from Manila to Java, then back to Manila once more to evacuate Francis Sayre, the US High Commissioner in the Philippines. *Atsutasan Maru* was merely the first in a long, tragic line of Japanese ships to vanish at the hands of the Silent Service - when it was all over, ninety percent of Japanese merchant shipping had been destroyed, the few surviving vessels restricted to night transits in waters where the submarines couldn't follow. *Swordfish*, sadly, would not see the day of reckoning: sometime between January 9th and February 15th, 1945, she was lost, possibly to a Japanese mine near Okinawa. Her exact resting place is unknown.

Swordfish - along with fifty-one other US submarines and three thousand, five hundred and five sailors, remain on patrol.

"To those whose contribution meant the loss of sons, brothers or husbands in this war, I pay my most humble respect and extend by deepest sympathy. As to the 374 officers and 3131 men of the Submarine Force who gave their lives in the winning of this war, I can assure you that they went down fighting and that their brothers who survived them took a grim toll of our savage enemy to avenge their deaths. May God rest their gallant souls."[98]

The **Last Ships Sunk By A Submarine** were, first, the Argentine Navy light cruiser ARA *General Belgrano,* sunk by Her Majesty's Ship *Conqueror* on May 2nd, 1982. The *Belgrano*, supporting Argentina's misbegotten efforts to conquer the Falkland Islands from Great Britain, was considered a serious threat to the RN task force sent to retake the islands - although she was a forty-seven year old former USN light cruiser, the USS *Phoenix*, if she had gotten loose among the lightly escorted supply and troop convoys she could have done devastating damage before being stopped. And since she had been built in an era where physical armor was a standard feature of all major

---

[98] VADM Charles Lockwood, CINCSUBPAC, speech in Cleveland, OH, 27 Oct 45 , in *United States Submarine Losses In World War II,* Naval History Division, Office of The Chief Of Naval Operations, Washington, DC, 1963

warships, the comparatively lightly armed British escorts would have had a hard time killing her. For their part, the Argentines had somehow convinced themselves that if their ships were outside the 200 nautical mile 'exclusion zone' around the islands - as was *Belgrano* - they would remain unmolested.

It was just one more Argentine miscalculation in a war that was full of them, but this one was particularly grim, and the foolishness of that assumption was made clear at just before 1600 hours local on May 2nd, 1982, a gray afternoon punctuated by 45-mph winds and eight-foot waves, with temperatures below freezing. The *Conqueror* had been tracking *Belgrano* for three days, and had requested permission to sink the old cruiser even though she was some distance outside the exclusion zone. Loosely escorted by the ex-US *Sumner* class destroyers ARA *Bouchard* and ARA *Pedra Buena*, the *Belgrano* never knew *Conqueror* was there.

And she never had a chance.

Conqueror fired three WWII-era Mk VIII torpedoes (her skipper was mistrustful of her state-of-the-art *Tigerfish* torpedoes), two of which hit the elderly cruiser. The first fish struck her midships, killing more than 250 sailors and taking out all the ship's electrical power, preventing any distress signals from being sent out. The second blew off her bow just forward of 'A' Turret, but amazingly caused no casualties. The third may have struck the *Bouchard* but failed to explode, a warning that does not seem to have sunk in with the destroyer's crew.

Belgrano's crew, considered an elite in the Argentine Navy, reacted with calm professionalism, getting the wounded on deck and trying to close off shattered bulkheads, but it was no use. *Belgrano* - a survivor of Pearl Harbor and Surigao Strait - was doomed. Less than twenty minutes later, the cruiser's skipper ordered the crew off. *Belgrano's* escorts, unable to see her in the poor weather and with their radars off, failed to notice their charge in agony for hours, and by the time they found her it was far too late. Three hundred and twenty five of her crew were lost, while seven hundred and seventy were rescued. Some, however, weren't rescued for another two days, spending what must have seemed an eternity in the vicious winter waters of the South Atlantic. A handful of rafts were later found containing only frozen corpses, and a forsaken few were never found at all.

Death In The South Atlantic: The nuclear attack submarine HMS *Conqueror* sinks ARA *General Belgrano* (ex-USS *Phoenix*) near the Falkland Islands on May 2nd, 1982. (Photo courtesy of Argentine Navy)

The Argentine Navy - including its sole carrier, the *Venticinco de Mayo* - made for port at flank speed and stayed there for the rest of the war. The Argentine government, embarrassed and furious, briefly had the nerve to call the sinking a war crime but that soon became the least of their worries as the Argentine people began to protest and eventually overthrow the government that had sent such a ship and crew so foolishly into harm's way.

There has been one more sinking on March 26, 2010, in the cold, murky waters of the Yellow Sea near the rocky shores of

Baengnyeong Island. The South Korean corvette *Cheonan* was on patrol at approximately 9:22 PM that night when something exploded near the *Cheonan's* stern. Whatever it was broke the corvette's back, and her stern came off at 9:27. The rest of the ship went under just three minutes later, taking 46 men with her. Whatever took down the proud little warship did it with utter and complete surprise - her crew didn't even have the warning the *Housatonic's* crew got, and nearby ships were in thorough confusion as to what might have happened.

Unto The Present Generation: *The Republic of Korea corvette Cheonan (PCC-772) is hit by a torpedo in the Yellow Sea and sunk with the loss of forty-six South Korean sailors on March 26th, 2010. Investigation finds parts of a North Korean torpedo in Cheonan's hull, but the final report very carefully avoids blaming the erratic nation and its leaders for the sinking. (Photo Xinhua News)*

Within minutes of course the North Koreans were the prime suspects in the sinking, and this led to a dilemma: if they did it, it was a violation of the 1953 armistice, and therefore an act of war. As straightforward as it seemed, that would also mean war with one of the most vicious, paranoiac, and - quite frankly - unbalanced regimes on the planet. The North Koreans replied to the suspicions with their usual berserk invective, denying any connection whatsoever with the sinking and threatening war if they were even suspected of being involved.

The Long Patrol

Matters were made only worse by the facts: after the wreck was lifted from the Yellow Sea, investigators found torpedo parts that bore a close resemblance to known North Korean torpedoes, plus the knowledge that at least two North Korean boats were in the vicinity. The damage to the hull was consistent with a torpedo strike. One can only imagine the dejected sadness of the investigators as they tried to find some logical reason for the loss besides a torpedo, but could only find solid evidence that said it was. The final report listed eleven - *eleven* - possible causes for the loss of the *Cheonan*, but the only one that fit all the facts was that she was executed by a North Korean submarine. Even then, however, the South refused to actually accuse the North, the report only stating that the ship was sunk by a torpedo, type and national origin unknown.

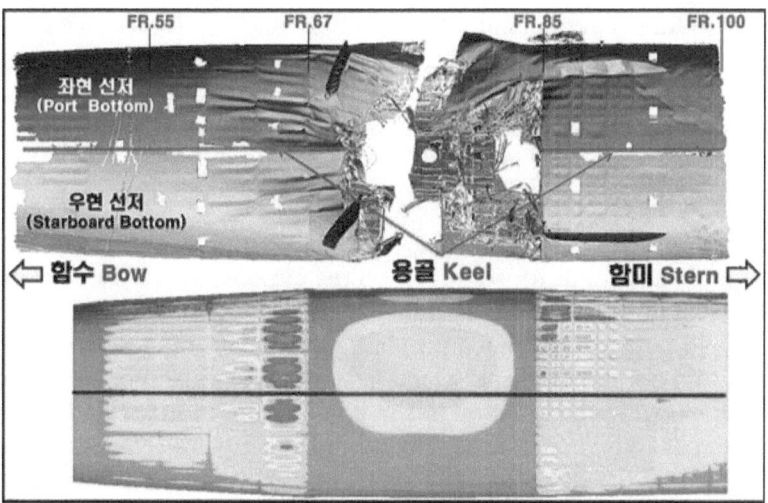

The Damage to the Cheonan; everybody knows the North Koreans did it. Admitting that is something else. Source Republic of Korea Navy

In the end, both of these incidents proved the point that George Dixon and the *Hunley* had made that cold February night in 1864: a determined submarine and a well-trained crew can strike without warning and without mercy....and will continue to do so.

Baxter Watson returned to New Orleans and died there not long after the end of the war.

CSS H.L. Hunley

William Alexander stayed in Mobile to the end, but the cannon he was working on was eventually tossed into some unknown river to avoid capture. He remained in Mobile for the rest of his life, which was a long and rewarding one. He became a highly successful engineer and fathered eight children, but he never forgot his friends who still waited for him in Charleston Harbor. He wrote several widely reprinted articles about the design and career of the *Hunley*, and for most of a century they were all we had to go on. He died, full of years, in 1914, just a few months before the beginning of the First World War. He is buried in Mobile's own Magnolia Cemetery.

The *CSS Saint Patrick* – the other submarine built in Mobile – could best be described as a cross between the *Hunley* and a *David*. It had a small but reasonably efficient steam engine and a retractable stack and pilothouse – then when it prepared for an attack, she could fully submerge and would switch over to human power. Apparently it worked quite well, and its size is believed to be approximately that of the *Hunley*, indicating that a steam plant wasn't impossible. On the other hand, given *Hunley's* design, it was probably quite impractical.

Saint Patrick's builder, John Halligan, took her out on several patrols but for some reason the circumstances were never quite right for her to go after a blockader. General Maury showed far more tolerance for Halligan's delays than Beauregard ever did for McClintock's, for it was almost six months from *Saint Patrick's* completion to the point where Maury finally lost his patience and ordered CSN Commander Eben Farrand to take control of the boat on January 24th, 1865. Farrand in turn assigned CSN Lieutenant John T. Walker to command her. Walker obeyed with alacrity and took her out just four days later after the blockader USS *Octorara*. Walker executed a brilliant attack, using every one of his advantages like a seasoned pro – and the spar torpedo failed to fire.

A surviving drawing of the event shows *Saint Patrick* – in this engraving bearing an amazing resemblance to the Walt Disney version of Jules Verne's *Nautilus* – scraping at high speed along *Octorara's* flank while a stunned Yankee bluejacket looks down in surprise, appearing as if he is about to step on

The Long Patrol

Saint Patrick's deck. Under fire from the *Octorara*, Walker brought her about and headed back to Mobile.

That was the end of *Saint Patrick's* combat career. She apparently served very briefly as a blockade-runner, foreshadowing the use of some huge German boats in this capacity early in the First World War. At some point prior to the capture of Mobile itself –anywhere from January 28th on to early April – *Saint Patrick* was scuttled; her exact whereabouts still a mystery. That makes two remarkable early submarines that lie missing somewhere in Mobile Bay; perhaps now that *Hunley* has returned from the other side the necessary resources may be brought to bear to bring *Pioneer II/American Diver* and *Saint Patrick* home.

And so we come back to Magnolia Cemetery, April 17th, 2004. For the crew of the *Hunley*, the Long Patrol is ending on this hot, humid day in South Carolina, and in the end that is what leads us to this not-quite time warp, where C-17 transports whine by on gas turbine wings, headed for Bosnia, Iraq, and parts unknown, where bright orange Coast Guard helicopters and gaudy TV news choppers whirr overhead…and the Confederate States Army wheels into formation for what may be its final pass-in-review.

It is said that LT Dixon had a keepsake – a $20 gold piece, given to him by his fiancé, Miss Queenie Bennett of Mobile. At the battle of Shiloh, Dixon took a round in the hip; the ball struck the gold piece squarely in its center forming it into the shape of a bell. There were witnesses to this minor miracle, so we know it happened. Dixon, for his part, kept the deformed coin with him for the rest of his life. Even with the eyewitness accounts, there was skepticism for years over the matter, but that was laid to rest one day in 2002, when a set of skeletal remains were delicately removed from where they had settled inside the *Hunley*.

Nestled inside them was a $20 gold piece, deformed into the shape of a bell.

One story that stays with us is that gold piece, given to a departing warrior by his love in a time when bravery and courage were watchwords of a life well lived, and how the honorable love of a Lady could urge a man on to feats of legend. We laugh now at such things. For all of our sophistication and knowledge, they are seemingly beyond us. We fight our wars in a cross between a science fiction movie and a reality television show that allows us no time for diversions. Yet on this Saturday morning, you hear those words in the crowd…the gold piece…the gold piece. And when a tall, dignified woman in black strides in with the crew families – all identified and confirmed by DNA – the words focus on her.

Her name is known, but it shall not be revealed here. She is Queenie Bennett's great-granddaughter, who has tried to stay out of the spotlight – for that matter, as have all the families. They do not look in any way noteworthy or distinctive – just the kind of people you would see at a family funeral. Only one stands out; an elderly lady of petite stature wearing a sash that indicates her as a member of a Confederate historical organization and is studded with a perfectly aligned row of pins showing her service. She is one of several crew family members who are at graveside when the service is done. A few inches behind me, someone asks her what ceremonies are next. She smiles and says, "What's next is that I intend to change into a comfortable pair of shoes."

South Carolina's politicians – enough of whom rest here at Magnolia, finally *sans phrase* – are of mixed emotions regarding the seven men and the thousands who have come to see them to their rest. The Confederate spirit died hard in the Palmetto state, and one can make a strong case that it has never died at all. Unfortunately for politicians who crave votes, the Confederacy is considered a very touchy subject. The matter of the Confederate naval flag – it was never the flag of the Confederacy itself – that flew over the State House in Columbia wiped out the career of one Governor and ignited a still-ongoing boycott by the NAACP.

The South Carolina State Museum - one of the true jewels of the state and its capitol - set up an exhibit explaining that most slaveholders actually tried to take care of their slaves, (if for no other reason than to preserve their investments) and set off a barrage of complaints that led to the re-wording and partial dismantlement of the exhibit. A Columbia restaurateur puts booklets in his business extolling the Confederacy, and is publicly pilloried for his efforts. Most try and avoid it, one notable exception being state Senator Glenn McConnell, who is to a great deal responsible for there even being a service this day and who is very nearly a guest of honor here.

But in the end, the politicians have pretty much run for cover on this one. Protests from black legislators scuttled plans for a lying-in-state at the State House, and there is no official delegation from Columbia here this morning. It is remarked upon by the crowd, but more on that later.

Magnolia is such a green oasis in the normally sere urban Charleston cityscape that anything black – expected though it may be here – catches the eye. At first, it registers on the conscious as the ubiquitous ravens that wheel and caw in ragged flocks throughout the countryside, but then you realize that these ebony harbingers move with a smooth precision that no avian can match.

They are the Widows – easily two hundred women and more, in full 1860s black mourning garb. They range from little girls not far removed from being toddlers to fresh-faced blonde teenagers, to stolid Charleston matrons and silver-trimmed grandmothers, each of them in the black laces and silks that were once a *de rigueur* display of mourning and sadness. They are not dressed simply to play a part here or gain attention – each and every one is there because they feel a deep and solid connection with the men in the varnished pine coffins and the women they left behind. They are here to bid a farewell for those who cannot, Confederate valkyries escorting the last of the faithful to Valhalla.

And of all the people there, they are the loudest and most vocal regarding the absence of a State delegation. One lady looks at me with a seriousness that belies her soccer-mom looks and says, quietly but firmly, "The Governor will be receiving a *lot* of letters about this." The rest nod in agreement and murmured assent, angry black angels among the emotionless white marble ones of Magnolia Cemetery.

CSS H.L. Hunley

The reenactors have arrived from all over the country, and for that matter, from outside the country as well; at least one British Confederate is across a still, quiet pond, helping prepare the artillery salute. It is probably not hyperbole to say that there have not been this many Confederate uniforms in one place in Charleston since the city was evacuated in early spring of 1865 – easily a light regiment's worth, carrying flags and guidons that identify them as Georgian, North Carolinian, Virginian, Tennesseean, Texan, Mississippian, South Carolinian, Floridian, Louisianan, Alabaman. And these are the organized units, each one with 'Volunteer Infantry' or 'Volunteer Artillery' in its name, reminding you that most Southerners fought because they wanted to, not because they had to in a hated draft. There eventually was a Confederate draft, but this does not seem to be the place to bring that point up.

There are hundreds more individuals in uniforms that reflect everything from the most humble Confederate militiaman to Generals and Admirals, with a smattering of sailors, Marines, and engineers thrown in. There are hundreds of civilian reenactors as well, dressed in flawlessly recreated period clothing. One – a university librarian – points out that there's really no way one can ever wear the real thing; it's simply not made for modern *homo sapiens* and if it does fit it's more an accident of nature than anything else. One other thing we must remember, she says – if you make your own, use cotton instead of polyester.

"It breathes," she smiles, pointing out that even in the eighty-degree temperature and almost physical humidity of Magnolia Cemetery, she's still holding up quite nicely, thank you.

And as the first units march into position around that little island in Magnolia Cemetery, she looks up at the sound of leather creaking and gear rattling, of hobnail boots tramping through hardpacked South Carolina dirt, of commands given and acknowledged, and a band a few hundred yards away playing 'Dixie' in notes that fade in and out as if on a radio that's not quite tuned into another reality. In a mixture of awe and sadness, she whispers, "The Confederacy has come back for a day."

It has. In these few square miles of South Carolina, it is not 2004 today – it is 1864, and I defy anyone to say otherwise. When I left home this morning, I was watching stories of 21st century technology being brought to bear on 12th century fanatics in a land whose civilization is tens of centuries old, but right here, right now, this is Somewhere Else. There is literally almost nothing to indicate that we are in the United States of America in 2004. There is not a single Star-Spangled Banner in sight, but there are battalions of the Stainless Banner, the true flag of the CSA – a white banner with the Saint Andrew's Cross in the upper left corner. On some of the more modern marble stones for CSA Civil War dead, there is no US seal, but rather the cross device of the Sons Of Confederate Veterans. And of course, the voices – my own Ohio twang submerged beneath tens of thousands of mellow Southern drawls.

Perhaps more disturbing though than the Twilight Zone aspects of this day are some of the snatches of conversation I hear through the crowd. The vast, overwhelming majority of people here today are here to honor nine brave men, to be a small part of a great history that will to a great extent wind up here today – not on a battlefield among the smoke and noise and cries of the wounded and dying, but instead in this quiet oasis where one can hide from the years. These are people who are truly touched today –they will remember it forever as did those who watched that first artillery round arc up in the air over Fort Sumter almost 143 years ago. It will be something remembered always and in awe, in whispers and hushed, reverent tones.

But there are some voices in the crowd that seem less interested in forgiving the differences than in celebrating them. One gentleman behind me – wearing a pin that proclaims his former leadership of a well-known Confederate heritage organization – is literally snarling as he points out to another man that he is damned glad there are no Yankee uniforms present this day.

"They had their ceremony," he growls. "This one is ours."

There are other rumblings of a similar nature that float in and out of hearing, some from the Distinguished Guests who line up

CSS H.L. Hunley

before us for photos. They are right, insofar as it goes. This is theirs. It will probably be the last Civil War funeral, and to that extent some will be able to claim that they were the Last Men Standing. It is difficult to believe that this far downrange it makes any difference, but to some it does.

The crew of the *H.L. Hunley* enters by stately procession into the road that circles the little island. They will in all likelihood be the last Confederate warriors to make the long trip from Charleston out to Magnolia. It is no longer a quiet country journey but instead a procession through battered public housing, past bustling container terminals, under busy bridges, through the seedy neighborhood that straddles Meeting Street. Nine reenactors, representing all the Confederate services and branches, carry crimson cushions ahead of them each with a medal draped across it – the SCV's Medal of Honor. The standards for its award are as high and stringent as those of the US Congressional Medal Of Honor, and it is more than fitting that these nine have earned it.

Seven men – six pallbearers and one detail commander, escort each crewman. The first thing that strikes you is that the coffins are so small, perhaps no more than about four and a half to five feet long. They are beautifully made from polished yellow pine, fitted together and finished so well that they appear almost as of a piece. The second thing that strikes you is the trim, impassive presence of a South Carolina Highway Patrolman at the left rear of each detail. They and the fatigue-and-flak-vest clad State Law Enforcement Division troops scattered around the cemetery are jarring reminders of where we really are. There is a genuine concern that someone will try and disrupt the ceremony – understandable enough, given some of the passions that have run high in this state. On the other hand there are a thousand or more reenactors here, many carrying a long rifle with a bayonet, a sword, or both. Anyone foolish enough to start something here could probably count on being carved up like an Easter ham live on South Carolina Public TV, which is broadcasting the ceremonies. That however does not always deter those who consider themselves Chosen, so the State has erred today on the side of caution.

The Long Patrol

Nevertheless, the State Troopers stay in position until each coffin is taken, gently and reverently, by the burial detail and laid in the long common grave that will be its resting place. One trooper stands at rigid attention as the coffin he escorts is taken for the last few feet of its journey. He is an example of the precision and professionalism that the State Troopers here show, and I think that in the heat and humidity it must be a challenge to keep looking that sharp, but he does it. Every movement is precise and sharp, every gesture smooth and respectful to the man he escorts.

I also wonder what he is thinking right now; for his is the only black face I have seen here all day.

A clergyman, resplendent in formal robes and flanked by two very serious young altar boys, reads the burial service. They are the words we have come to know so very well through other, sadder, more personal experiences – a reminder that in the end we all return to the Earth our Lord made us from, and that even then we but sleep; confident in resurrection. A group of Masons, almost incongruous in tuxedos, bow ties, and aprons, appears momentarily as LT Dixon is laid to his rest – Dixon was a Mason, and his fellows, known for their fraternity and fellowship, are determined to give their brother a proper farewell.

The others approach, halt, and are guided into the grave. There is a silence that one does not associate with crowds this large. The only sounds are distorted echoes of a news helicopter, and the precise tread of a few more companies of troops moving into position. Even the cameras are muffled, their whirrs and beeps curiously missing at this final moment. Perhaps the technology that enables us to preserve moments like this for all time is, in the end, not as good as the pictures we will keep in our memories until the day we die. We will take the boundless pictures of our mind, with every color and sound and smell not only there forever, but enhanced, over the cold, brittle feel of a CD with a computer program on it.

It is done now, almost without warning. A bugler steps up and plays a quick tattoo. He stumbles slightly over some of the notes, but no one chastises him, for his effort comes from the heart. Then, after a heartbeat's pause comes the melody written by a Civil War general, one meant to signify the end of the duty day but instead one we have come to associate with sacrifice, tragedy, and lost causes...

"*Day is done...*"

We know what to expect now – first, a fifty-gun salute by the troops gathered around the gravesite. I am used to the flat, sharp crack of M-16s from twenty years in the Air Force but I am not prepared for the deep, growling bark from fifty muzzle-loading rifles, and I know now why grown men quailed at that sound. Flame lances out literally three feet in front of them and Spanish moss is shredded into streamers and confetti, drifting down onto the stones that mark LT Payne's crew and Captain Hunley's lost souls.

Then, the finale – fifty cannon, lined up wheel to wheel on the other side of the pond that sits to our right. The guns – at least one an actual Civil War veteran from the Battle of Sharpsburg (or Antietam, depending on which side of the fight your ancestry placed you) – thunder loose with a chest-thumping *thud*, and curiously there is little echo before two and one half seconds later the next one lets go, and so on and so on. As the guns fire down the line, the sound grows more and more distant and oddly higher pitched until the last few seem to be coming from somewhere else entirely – the far end of the cemetery, perhaps, or another time entirely.

With the last cannon shot fading away, it is done. Our friend the librarian is here with one of the Artillery companies, and with a gentle smile tells us that she has to return to her unit to help them limber up the guns. She is thrilled to have seen something like this, but just a little disappointed – "Normally," she says, "*I'm* on one of the gun crews." We say our goodbyes and nice-to-have-met-yous, and she disappears into the crowd, hoop skirt swaying, one hand holding a high-tech digital camera and an antique lace folding fan.

In a very calm, dignified wave, the thousands edge close to the gravesite. They part to allow the families another quick glimpse, then close up again. In a few short minutes, I am at the graveside.

Hunley's crew is in their final military formation, one they shall maintain until the Last Trump releases them and the other faithful of Magnolia Cemetery. The coffins are lined up with precision, brass plates on each lid identifying the man who rests inside. Each one has a stunningly pink Confederate Rose atop it, odd in the martial tone of this day but a thankfully human touch. There are a few grains of South Carolina sand atop each coffin, as if reminding its resident that in a few minutes the darkness shall return.

"...In their final military formation, one they shall maintain until the Last Trump releases them..." *Requiescat im Pacem,* April 17th, 2004. (Photo by Theodore Leverett)

My attention is drawn to a hollow thump and I see dirt cascading down onto a coffin. For a moment I think that someone has slipped and is falling down into the grave, but instead it is one man – then two, then others – quietly picking up a handful of soil and almost reverently tossing it into the grave. The State Troopers on the other side of the grave simply nod, and everyone on either side of me begins to do it as well. I bend

CSS H.L. Hunley

down and grasp a small handful, rough and dry in my hand. Just a flick of the wrist – and it sprays out and down, a roostertail suspended in the air for a moment before the faint breeze blows it away. In a way I cannot explain, I have connected with these seven, dead almost a century before my own birth.

To me, they have never been merely Confederate or Federal, but rather Americans who took their beliefs and abilities to that final limit that we rarely see any more. I have read their story over and over again; I have seen their ship in its berth within a high-tech lab, I have followed the debate over their honor, but nothing has closed that gulf between us as much as the act of helping – in a very, very small way – to lay them to their rest. I dust my hands off and look for my partner to begin the walk back to Meeting Street.

We walk down the narrow brick-and-dirt path that goes back to Magnolia's modest gate. There are other names on other stones to recognize here – Commodore Duncan Ingraham, the CSN officer who sent Dixon and Alexander to the *Indian Chief* to find a crew. Robert Barnwell Rhett, the firebrand whose newspaper the Charleston *Mercury* did so much to whip up the passions that led so many to their final rest around him.

As we near the gate, the infantry companies come tramping back up the path behind us and across the bridge that spans the pond to our right. They are military, disciplined – they are professional, even though many have never served in the armed forces. There is no sound other than the commands of the officers and sergeants carefully threading them through the cemetery's lanes. I have never seen military units literally on the march – my experience has been on trucks and aircraft, squared-off armored personnel carriers, but this is something utterly alien to me, like a history book that has suddenly come to life. I have seen it in my mind's eye since I was a child, but it's here in front of me now.

A voice that can only belong to a First Sergeant roars out, "Make way for soldiers!" and the crowds part with a speed and precision one does not normally associate with civilians. Up the path comes a company's worth of infantry, probably far better equipped, shod, uniformed, or fed than the real thing ever was. The clatter of hobnailed boots on brick fills the air around us, and they stand as tall and as straight as any I have ever seen.

The Crew Of the H.L. Hunley Laid To Rest. (Photo by Theodore Leverett)

The Confederacy never had a victory parade, of course. They were never able to pull troops off the line to come home to Richmond or Atlanta or Charleston to enjoy just one brief moment of acclaim or gratitude, and when they finally did return to their homes it was as ragged, individual survivors. Today though the flag of the Confederacy flies here proudly for the first time in one hundred and thirty-nine years, at the head of a column of Confederate infantry – and it is as much a victory parade as the one that finally tramped down dusty Pennsylvania Avenue in Washington when the Civil War ended, or the one eighty-one years later after the Second World War.

There are the better part of one hundred thousand people here at Magnolia on this humid April afternoon. It's unlikely there were ever that many in once place cheering for the Confederate States Army during the Civil War, but there are

CSS H.L. Hunley

now. This is their victory parade, the one that the Confederate States Army never had for Fort Sumter, Manassas, Chancellorsville, Fredericksburg, and all the other times they sent Federal forces reeling back, fleeing before that unearthly battle cry which has come down to us as the Rebel yell.

One hundred and thirty-nine years after the last Stainless Banner came down[99], more than half a century after the last confirmed survivors went to their rest[100], the Confederate Army is marching one last time with every flag and guidon flying and with cheering crowds beside them as they march back into history. And as they come past me, only a few steps from the gate, it is a reflex from twenty-six years into my own past that snaps me to attention. Many are saluting the Stainless Banner. I cannot; there is a hesitation in me that restrains me until it passes, but I have no qualms about standing tall before the men who are literally the heirs of Lee's Legions – and those of Jackson, and Longstreet, and even Beauregard and all the rest.

Perhaps somewhere else, April 12th – Secession Day - is a holiday for half a continent, and legions in Butternut and Gray march down paved canyons of skyscrapers while historians discuss in dry academic tomes discuss what would have happened had the South lost. It is but a fantasy – but it is probably the most common and most discussed one in American history, and today we are seeing it. Not on a movie screen or

---

[99] The last ground units surrendered on May 10th, but some units of the CSN didn't get the word for months – most notably the raider *Shenandoah*. She killed a Union whaler on June 28th, 1865, and wasn't advised of the surrender until a British frigate caught up with her on August 2nd., and even then her crew was at best reluctant to face the truth. (See *Last Flag Down,: The Epic Journey Of the Last Confederate Warship*, by John Baldwin and Ron Powers,) Putting into Liverpool, her commander didn't haul down her flag until the crew was paroled on November 8, 1865. Some Confederates, most notably Matthew Maury, fled to Mexico in an attempt to found a Confederate colony there. They managed to hold on – officially – until 1868, and there is still a sizable Confederate colony in Brazil - MJK

[100] They were, respectively, Pleasant Crump of the 10th Alabama who died in 1951 and James Woolson, a Federal drummer boy who died in 1956. The last Civil War widow – Alberta Martin of Alabama, whose husband had fought with the 4th Alabama and married her when he was 81 and she 21 – died on June 1st, 2004, just a few short weeks after the *Hunley's* crew was laid to rest. The books were not then officially closed on the War Between The States until one hundred and thirty nine years after the fighting stopped - MJK

some computer generated spectacular on the History Channel but literally close enough to reach out and touch. And the crowd here is on their feet, whooping and cheering the soldiers on as they head for the gate.

As the first company clears the gate, it happens – not a command given, not a word spoken, but from just a few feet in front of me comes that yell. It sounds the way some huge wave *looks* – swelling up, literally rising over your head and curling up as it comes down on top of you, your hair standing on end and goose bumps running up and down your arms. This was the sound that terrified McDowell's men at Bull Run, the sound that drowned out the bugles on a hundred of Stuart's charges, the sound that echoed up ahead of a doomed division at Pickett's Meadow.

And again, without a command given, as soon as the last man of each unit clears Magnolia's gate, they give that yell again and again and again. Children, from toddlers to teens, are running out behind the units now and chasing them, keeping up with them, talking to them. More than a few young ladies are flirting with and waving at the soldiers as they head up towards Meeting Street and then turn off towards their assembly points, disappearing in haze and dust.

It strikes me how much it must have looked like this as the units marched out of another Charleston in another time, banners flying, fair Southern maidens cheering them on, and children looking up admiringly. It isn't the 21st century here today, not at all. As I turn towards where my car is parked, I walk through formations of soldiers, sailors, Marines, blockade running captains and Southern belles, all talking about the same things they would have been discussing a hundred and forty years ago. Another time, a different world.

Fifteen minutes and one hundred and forty years later, we start the long drive home.

M.J.K.

October 17th, 2012

Acknowledgements

No work like this can be possible without the aid and assistance of many, many people, and this book is no exception.

I am indebted beyond words to everyone whose help made it possible. Mr. Joe Long of the Confederate Relic Room at the State Museum in Columbia, SC, was kind enough to open their archives to me, giving me access to the *Official Records* and many other works that were invaluable in putting this together. My father, Donald Kozlowski, moved from a career at NASA to reviewing my data on a Civil War era submarine and his insights and opinions were priceless. Bob Dedmon – a man I have been proud to call my best friend and brother for almost three decades – offered patient encouragement every step of the way and made sure I never gave up on it. A group of wonderful friends, some whom I have not yet met in person – Stuart Slade, Jeff Thomas, Andy Pico, Joe Czarnecki, AT Mahan, Tony Evans, the Scholarian, Allen Hazen, Case, SunniBrook, Karl Newman, Kevin, The Duchess, Dennis, p620346, the Argus, and others who shall go by the collective name of the Board, generously donated their priceless historic and technical knowledge.

Another member of the Board, Theodore Leverett, accompanied me to the *Hunley* memorial ceremony in April 2004, and took some of the wonderful pictures that appear in this work. I have learned and laughed with these friends, and I will never be able to thank them enough for their help, advice, and encouragement. My son Michael was with me through every moment of the original research and writing, and took as much joy in all of this as I did. My wife Melissa – a Lady in the finest sense of the word – gave me the encouragement needed to get the final revisions done and on their way.

I also want to thank Debbie Ciannella, who, many years ago, saw something in my work that others didn't and worked long and hard to try and make it a success. I am forever indebted to her for it.

Of course, any and all errors are mine and mine alone.

MJK October 2012

References and Bibliography

<u>Hard Copy</u>

Augustin, George. *History Of Yellow Fever*, Searcy and Pfaff, New Orleans, LA, 1909

Burton, E. Milby. *The Siege of Charleston 1861-1865*, University of South Carolina Press, Columbia, SC, 1970

Carnes, Mark C., ed. *American National Biography, Supplement 2*, Oxford University Press, New York, New York, 1998

Catton, Bruce. *The American Heritage Picture History Of The Civil War*, Random House, New York, 1960.

Cussler, Dr. Clive. *The Sea Hunters*, Pocket Star Books, New York, 1996

Cussler, Dr. Clive. *The Sea Hunters II*, Berkley Books, New York, 2002

Government Printing Office. *Official Records Of The Union And Confederate Navies In The War Of The Rebellion*, Government Printing Office, Washington DC, 1903

Government Printing Office. *War Of The Rebellion: A Compilation Of The Official Records Of The Union And Confederate Armies,* Government Printing Office, Washington, DC, 1890

Hicks, Brian, and Kropf, Schuyler. *Raising The Hunley: The Remarkable History And Recovery Of The Lost Confederate Submarine.* Ballantine, New York, 2002.

Konstam, Angus. Confederate Submarines and Torpedo Vessels 1861-65, Osprey Publishing, Oxford, UK, 2004

Maury, Dabney H. *Recollections Of A Virginian In The Mexican, Indian, And Civil Wars,* Charles Scribner's Sons, New York, 1894. (Edition cited is from the online resources of the University Of North Carolina at Chapel Hill, http://docsouth.unc.edu/maury/maury.html)

McPherson, James M. *Battle Cry Of Freedom*, Oxford Univ.

Press, New York, 1988.

Naval History Division, Department of the Navy. *Civil War Naval Chronology 1861-1865*, US Government Printing Office Washington, DC 1971

Peltzer, John. 'The Union's Mission to Relieve Fort Sumter', *America's Civil War Magazine*, September 1997

Ragan, Mark. *The Hunley: Submarines, Sacrifice & Success In The Civil War*, Narwhal Press, Charleston, SC, 1999.

Steers, Edward. *Blood on the Moon: The Assassination of Abraham Lincoln*, The University Press of Kentucky, Lexington, KY, 2000.

Wills, Rich. *The H.L. Hunley In Historical Context*, Naval Historical Center, Washington, DC, 2001

Monk, John. *"Government By Stealth: How Senator Steers Sub Under Radar"*, The State (Columbia, SC), May 6, 2006

Internet Sources

Centers for Disease Control and Prevention, Atlanta, GA. *Smallpox: Disease, Prevention, and Intervention* (www.bt.cdc.gov/agent/smallpox/training/overview), 2012

www.lsm.crt.state.la.us/cabildo/cab8a.htm

www.militarymuseum.org/CAandCW2.html

www.thehunley.com

www.hunley.org

http://www.ussvi.org/mem/mstrysub.htm

http://xroads.virginia.edu/~UG97/monument/maurybio.html

www.numa.org

Organizations

Department Of The Navy - Naval Historical Center
805 Kidder Breese SE
Washington Navy Yard
Washington DC 20374-5060

CSS H.L. Hunley

The Museum Of Charleston
360 Meeting Street
Charleston SC 29403

Fort Moultrie National Monument
1214 Middle Street
Sullivan's Island, SC 29482

APPENDICES

APPENDIX 1: CSS Pioneer

(From the United States Navy Historical Center, http://www.history.navy.mil/danfs/cfa8/pioneer.htm)

Pioneer (SS)

Pioneer, a privateer two-man submarine, was begun in New Orleans in 1861 to meet the menace of the United States steamers *New London* and *Calhoun* on Lake Pontchartrain. She was completed in early 1862, having been constructed from quarter-inch riveted iron plates that had been cut from old boilers. Some reports indicate that *Pioneer* was built in the Leeds Foundry but her principal inventor, J. R. McClintock, stated that she was built in his Machine Shop at 21 Front Levee Street, where, in partnership with B. Watson, he manufactured steam gages and turned out "minnie balls" on a high-speed machine of his own invention.

According to a post Civil War interview with McClintock, *Pioneer* was 30 feet long, of which a 10-foot midships section was cylindrical. From either end of the cylinder was a tapered section that gave her conical ends and resulted in a kind of "cigar shape." There was a conning tower with manholes in the top and small windows of circular glass in her sides. One man propelled the submarine by turning the manual crank of the screw and her iron ballast keel, detachable from the inside, was so heavy that it barely enabled the submarine to float on the surface with the conning tower awash. She was equipped with diving planes and was armed by a clockwork torpedo, carried on top of the submarine, and intended to be screwed into the bottom of the enemy's ship by gimlet-pointed screws of tempered steel. Actual inside measurements of *Pioneer* made on the spot by W. M. Robinson in 1926, were reported by him to be: length of 20 feet; maximum inside width of 3 feet, 2 inches; and a maximum depth of 6 feet.

There is little clear evidence on the operations of *Pioneer.*

The Long Patrol

F. H. Hatch granted her a commission as a privateer on 12 March 1862 and the application for a letter of marque was forwarded to Richmond on 1 April 1862. Her register of commission listed J. K. Scott as *Pioneer's* commander, and he described her as 34 feet in length, 4 feet breadth, 4 feet deep; measuring about 4 tons, with round conical ends and painted black. Her part owners were identified as J. K. Scott, R. F. Barrow (brother-in-law of H. L. Hunley), B. Watson and J. R. McClintock. A surety bond of $5,000 was put up by H. L. Hunley and his lifelong friend and college classmate, Henry J. Leovy who was then a New Orleans attorney of the law firm of Ogden and Leovy.

The application for letter of marque for *Pioneer* called "for authority to cruise the high seas, bay, rivers, estuaries, etc., in the name of the government, and aid said Government by the destruction or capture of any and all vessels opposed to or at war with the said Confederate States, and to aid in repelling its enemies." In an interview after the Civil War, McClintock stated *Pioneer* made several descents in Lake Pontchartrain and succeeded in destroying a small schooner and several rafts during experiments. Before she could attack a Union ship, Farragut captured New Orleans and she was sunk to prevent her from falling into Federal hands.

Pioneer was recovered long after the Civil War and removed to Camp Nicholls, the Louisiana Home for Confederate Soldiers. On 24 April 1957 she was transferred to her present site in the Presbytere Arcade, Louisiana State Museum, New Orleans, La. She was the forerunner of two other submarines that were built at Mobile, Ala., one the unnamed submarine sometimes called *"Pioneer II"* (q.v.) and *H. L. Hunley* (q.v.).

(NOTE: It has since been determined that this vessel is not the *Pioneer,* and its actual identity is unclear – MJK.)

CSS H.L. Hunley

APPENDIX 2: Intelligent Whale

(From the United States Navy Historical Center, http://www.history.navy.mil/danfs/i2/intellig.htm)

Intelligent Whale

(Submarine: weight 4,000 pounds (estimated); l. 28'8"; b. 7'; dph. 9'; s. 4 k.; cpl. 6 to 13)

Intelligent Whale, an experimental iron-hulled submarine, was built at Newark, New Jersey, to the design of Scovel S. Merriam, who entered into an agreement on 2 November 1863 with Augustus Price and Cornelius S. Bushnell. In April 1864, the American Submarine Company was formed, taking over the interests of Bushnell and Price. Years of litigation, however, followed, while those who built the boat apparently encountered "great difficulty...in getting a crew to man her for her first test in Newark Bay."

Intelligent Whale could be submerged by filling compartments with water, and then expelling the water by pumps and compressed air. It was estimated that the supply of compressed air inside could allow the boat to stay submerged for about 10 hours. Thirteen crewmen could be accommodated, but only six were needed to make her operational, motive power being furnished by a part of the crew cranking, attaining a speed of about four knots. General Thomas William Sweeny, a colorful decorated veteran of the Mexican War and Civil War (and who would participate in the Fenian Invasion of Canada later in the year), and two other men, tested the boat in April 1866. They submerged her in 16 feet of water, and Sweeney, clad in a diver's suit, emerged through a hole in the bottom, placed a charge under a scow, and reentered the submarine. When Intelligent Whale was a safe distance away, Sweeny exploded the charge by a lanyard and a friction primer, blowing the scow to pieces. Ultimately, at the end of the period of litigation, however, a sheriff's sale disposed of the boat. With the establishment of the title in court, the boat ultimately belonging to an "Abe" Halstead, the submarine was sold on 29 October 1869 to the Navy Department, for the following terms: $12,500 to be paid upon making and signing the agreement, $12,500 upon completion of the successful experiment, and $25,000 for all "secrets and inventions" connected with the craft.

A trial of Intelligent Whale occurred at the New York Navy Yard, Brooklyn, N.Y., three years later. "After sinking the boat, it was found the opening on top was leaking through defective packing," reported the Army and Navy Journal of 21 September 1872, "and after remaining under water a short time, the leak was so bad it was found expedient to raise her, but in doing so she caught under the derrick, and signals were sent to those on board [the derrick] to hoist the boat out, which they did. In the meantime, those on terra firma were excited by the fear that some serious mishap would occur to the persons in the torpedo-boat [one of whom was, apparently, Halstead himself], but after having been under the water sometime in the same spot, nor having traveled or accomplished anything, the boat was got out, and found nearly half full of water, her navigators unhurt, but we imagine, considerably frightened…"

Upon the conclusion of the unsuccessful test, the Navy refused further payments and abandoned the project. Some credit Intelligent Whale with inspiring John Holland to pursue submarine design.

Displayed for a time at the New York Navy Yard and then the Washington Navy Yard, Intelligent Whale as of this writing (October 2006) is on display at the National Guard Militia Museum of New Jersey in Sea Girt, New Jersey.

CSS H.L. Hunley

APPENDIX 3: Pioneer II

(From the United States Navy Historical Center, http://www.history.navy.mil/danfs/cfa8/pioneer_ii.html)

Pioneer II *(Author's Note: This is the first Mobile boat, known both as Pioneer II and/or American Diver – MJK)*

(SS: l. 36'; b. 3'; dph. 4'; cpl. 5; s. 2.5 k.; a. clockwork torpedo)

An unnamed submarine sometimes called *"Pioneer II"* was built during 1863 in the machine shop of Park and Lyons, Mobile, Ala., on plans said to have been furnished by H. L. Hunley, B. Watson and J. R. McClintock. Her principal builder was probably W. A. Alexander who claimed this distinction after close of the Civil War, stating that he was a Confederate Army Engineer of Company B, 21st Alabama Volunteer Regiment, CSA. He also stated he was assisted by Lt. G. E. Dixon, Company A, 21st Alabama Volunteer Regiment, who had also been detailed to do work in the machine shop of Park and Lyons.

H. L. Hunley has left record that he provided the "entire means" for this five-man submarine and McClintock stated that much money was spent in an unsuccessful attempt to power with an electro-magnetic engine. He afterwards fitted cranks to turn the propeller by hand, working four men at a time, but was unable to get a speed sufficient to make the submarine of service against Union ships blockading Mobile. In a letter to M. F. Maury in 1868, McClintock gave her dimensions as 36 feet long, 3 feet wide, 4 feet deep, with 12 feet of each end being tapered to facilitate underwater movement. She was towed off Fort Morgan to be manned for an attack on the Federal Fleet but foul weather and rough seas swamped her, without any loss of life.

"Pioneer II" was probably the submarine described by a Confederate deserter on 26 February 1863 to the Senior Officer of the Federal Blockade off Mobile: "On or about the 14th, an

infernal machine, consisting of a submarine boat, propelled by a screw which is turned by hand, capable of holding 5 persons, and having a torpedo which was to be attached to the bottom of a vessel and exploded by means of clockwork, left Fort Morgan at 8 p.m. in charge of a Frenchman who invented it. The intention was to come up at Sand Island, get the bearing and distance of the nearest vessel, dive under again and operate upon her; but on emerging they found themselves so far outside the island and in so strong a current (setting out) that they were forced to cut the torpedo adrift and make the best of their way back." She was second in a line of three submarines that included *Pioneer* (q.v.) and *H. L. Hunley* (q.v.).

APPENDIX 4: The Galena Letter

(From http://www.rootsweb.com/~iljodavi/news/3131864.htm)

GALENA DAILY GAZETTE
OLD SERIES---VOL. XVI, NO. 144
NEW SERIES---VOL. I, NO. 25
Monday morning, March 13, 1864

Page 2 Col. #2

THE LOSS OF THE HOUSATONIC

--"A Naval Officer's Account of the Affair -- A terrible Agency in Naval Warfare"

"As a history of the recent disaster of the U.S. steamer Housatonic may be of interest to many of your readers, I will attempt a brief statement of facts.

On the evening of February 17th, the Housatonic was anchored outside the bar, two and a half miles from Breach Inlet battery, and five miles and three-fifths from the ruins of Sumter-- her usual station on the blockade. There was but little wind or sea, the sky was cloudless and the moon shining brightly. A slight mist rested on the water, not sufficient, however, to prevent our discerning other vessels on the blockade two or three miles away. The usual lookouts were stationed on the forecastle, in the gangway and on the quarterdeck.

At about 8:45 of the first watch, the officer of the deck discovered, while looking in the direction of Breach Inlet battery, a slight disturbance of the water, like that produced by a porpoise. At the time it appeared to be about one hundred yards distant and a-beam. The Quartermaster examined it with his glass, and pronounced it a school of fish. As it was evidently nearing the ship, orders were at once given to slip the chain, beat to quarters, and call the Captain. Just after issuing these orders, the Master's Mate from the forecastle reported the suspicious appearance to the officer in charge. The officers and men were

The Long Patrol

promptly on deck, but by this time the submarine machine was so near us that its form and the phosphorescent light produced by its motion through the water were plainly visible. At the call to quarters it had stopped, or nearly so, and then moved towards the stern of the vessel, probably to avoid our broadside guns. When the Captain reached our deck, it was on the starboard quarter, and so near us that all attempts to train a gun it were futile. Several shots were fired into it form revolvers and rifles; it also received two charges of buckshot from the Captain's gun.

The chain had been slipped and the engines had just begun to move, when the crash came, throwing timbers and splinters into the air, and apparently blowing off the entire stern of the vessel. This was immediately followed by a fearful rushing of the water, the rolling out of a dense, black smoke from the stack, and the settling of the vessel.

Orders were at once given to clear away the boats, and the men sprang to work with a will. But we were filling too rapidly. The ship gave a lurch to port and all the boats on that side were swamped. Many men and officers jumped overboard and clung to such portions of the wreak as came within reach, while others sought safety in the rigging and tops. Fortunately we were in but twenty-eight feet of water, and two of the boats on the starboard side were lowered. Most of those who had jumped overboard were either picked up or swam back to the wreck. The two boats then pulled for the Canandaigus, one and a half miles distant. Assistance was promptly rendered by that vessel to those remaining on the wreck.

It was the opinion of all who saw the strange craft, that it was very nearly or entirely under water, that there was no smoke-stack, that it was from twenty to thirty feet in length, and that it was noiseless in her motion through the water. It was not seen after the explosion. The ship was struck on the starboard side abaft the mizzen-mast. The force of the explosion seems to have been mainly upward. A piece ten feet square was blown out of her quarterdeck, all the beams and carlines being broken transversely across. The heavy spanker boom was broken in its thickest part, and the water for some distance was white with splinters of oak and pine. Probably no more than one minute

CSS H.L. Hunley

elapsed from the time the torpedo was first seen, until we were struck, and not over three or four minutes could have passed between the explosion and the sinking of the ship. Had we been struck in any other part, or before the alarm had been given the loss of life would have been much greater.

The Housatonic was a steam-sloop, with a tonnage of 1,240, and she carried a battery of thirteen guns. She was completed about eighteen months ago, and has been in the blockade ever since. She is the first vessel destroyed by a contrivance of this character, and this fact gives to this lamentable affair a significance which it would not otherwise possess. Deserters tell us that there are other machines of this kind in the harbor, ready to come out, and that several more are in process of construction. The country cannot attend too earnestly to the dangers which threaten our blockading fleets, and the gunboats and streamers on the Southern rivers."

-----Off Charleston, Feb. 22, 1864

The Long Patrol

CSS H.L. Hunley

George Washington	381'7"	33'	6709 tons submerged	Nuclear	20 kts surfaced 25 kts+ submerged	12 Officers, 100 Enlisted	6x bow torpedo tubes 16x *Polaris* missile tubes	1959
Sturgeon	292'	32'	4309 tons submerged	Nuclear	25 kts	14 Officers, 95 Enlisted	4x bow torpedo tubes	1967
Projekt 971 (USSR)	361'	44'	8450 tons submerged	Nuclear	10 kts surfaced 35 kts submerged	31 Officer, 31 Enlisted	8x bow torpedo tubes 1x SAM launcher	1984
Los Angeles	360'	33'	6927 tons submerged	Nuclear	30 kts + submerged	13 Officers, 116 Enlisted	4x bow torpedo tubes	1976
Projekt 941 (USSR)	574'2"	75'6"	48,000 tons submerged	Nuclear	22 kts surfaced 27 kts submerged	50 Officers, 100 Enlisted	6x bow torpedo tubes 20x R-39 missile tubes	1981
Ohio	560'	42'	18,750 tons submerged	Nuclear	25 kts submerged	15 Officers, 140 Enlisted	4x bow torpedo tubes 16x *Trident* missile tubes	1981
Seawolf	353'	40'	9137 tons submerged	Nuclear	35 kts submerged	12 Officers, 121 Enlisted	8x bow torpedo tubes	1997
Virginia	377'	34'	7800 tons submerged	Nuclear	25+ kts submerged	11 Officers, 102 Enlisted	4x bow torpedo tubes 12 VLS tubes	2004

The Long Patrol

George Washington	381'7"	33'	6709 tons submerged	Nuclear	20 kts surfaced 25 kts+ submerged	12 Officers, 100 Enlisted	6x bow torpedo tubes 16x *Polaris* missile tubes	1959
Sturgeon	292'	32'	4309 tons submerged	Nuclear	25 kts	14 Officers, 95 Enlisted	4x bow torpedo tubes	1967
Projekt 971 (USSR)	361'	44'	8450 tons submerged	Nuclear	10 kts surfaced 35 kts submerged	31 Officer, 31 Enlisted	8x bow torpedo tubes 1x SAM launcher	1984
Los Angeles	360'	33'	6927 tons submerged	Nuclear	30 kts + submerged	13 Officers, 116 Enlisted	4x bow torpedo tubes	1976
Projekt 941 (USSR)	574'2"	75'6"	48,000 tons submerged	Nuclear	22 kts surfaced 27 kts submerged	50 Officers, 100 Enlisted	6x bow torpedo tubes 20x R-39 missile tubes	1981
Ohio	560'	42'	18,750 tons submerged	Nuclear	25 kts submerged	15 Officers, 140 Enlisted	4x bow torpedo tubes 16x *Trident* missile tubes	1981
Seawolf	353'	40'	9137 tons submerged	Nuclear	35 kts submerged	12 Officers, 121 Enlisted	8x bow torpedo tubes	1997
Virginia	377'	34'	7800 tons submerged	Nuclear	25+ kts submerged	11 Officers, 102 Enlisted	4x bow torpedo tubes 12 VLS tubes	2004

CSS H.L. Hunley

The Long Patrol

www.ingramcontent.com/pod-product-compliance
Lightning Source LLC
Chambersburg PA
CBHW030136170426
43199CB00008B/91